ETHICS AND CRITICAL CARE MEDICINE

PHILOSOPHY AND MEDICINE

Editors:

H. TRISTRAM ENGELHARDT, JR.

The Center for Ethics, Medicine and Public Issues,
Baylor College of Medicine, Houston, Texas, U.S.A.

STUART F. SPICKER

University of Connecticut, School of Medicine,
Farmington, Connecticut, U.S.A.

VOLUME 19

ETHICS AND CRITICAL CARE MEDICINE

Edited by

JOHN C. MOSKOP

and

LORETTA KOPELMAN

East Carolina University School of Medicine, Greenville, N.C., U.S.A.

D. REIDEL PUBLISHING COMPANY

A MEMBER OF THE KLUWER  ACADEMIC PUBLISHERS GROUP

DORDRECHT / BOSTON / LANCASTER / TOKYO

Library of Congress Cataloging in Publication Data

Main entry under title:

Ethics and critical care medicine.

(Philosophy and medicine ; v. 19)

Based on papers presented at a symposium held at East Carolina University School of Medicine in Greenville, N.C. on Mar. 17–19, 1983; sponsored by the East Carolina University School of Medicine and others.

Includes bibliographies and index.

1. Critical care medicine—Moral and ethical aspects—Congresses. 2. Critical care medicine—Social aspects—Congresses. 3. Long-term care of the sick—Moral and ethical aspects—Congresses. 4. Triage (Medicine)—Moral and ethical aspects—Congresses. 5. Medical ethics—Congresses. I. Moskop, John C., 1951– . II. Kopelman, Loretta M. III. East Carolina University. School of Medicine. IV. Series. [DNLM: 1. Ethics, Medical—Congresses. 2. Critical Care—Congresses. W3 PH609 v.19 / WB 105 E84 1983]

R725.5.E84 1985 174′.2 84–24796

ISBN 90–277–1820–2

Published by D. Reidel Publishing Company,
P.O. Box 17, 3300 AA Dordrecht, Holland

Sold and distributed in the U.S.A. and Canada
by Kluwer Academic Publishers,
190 Old Derby Street, Hingham, MA 02043, U.S.A.

In all other countries, sold and distributed
by Kluwer Academic Publishers Group,
P.O. Box 322, 3300 AH Dordrecht, Holland.

Printed in The Netherlands

For Daniel Moskop
For Elizabeth and William Kopelman

TABLE OF CONTENTS

ACKNOWLEDGEMENTS ix

INTRODUCTION xi

WARREN THOMAS REICH / A Movable Medical Crisis 1

WARREN THOMAS REICH / Moral Absurdities in Critical Care Medicine: Commentary on a Parable 11

H. TRISTRAM ENGELHARDT, JR. / Moral Tensions in Critical Care Medicine: "Absurdities" as Indications of Finitude 23

WARREN THOMAS REICH / "Conceptual Construals" vs. Moral Experience: A Rejoinder 35

JAY KATZ / Can Principles Survive in Situations of Critical Care? 41

STUART F. SPICKER / Coercion, Conversation and the Casuist: A Reply to Jay Katz 69

LORETTA M. KOPELMAN / Justice and the Hippocratic Tradition of Acting for the Good of the Sick 79

JAMES M. PERRIN / Clinical Ethics and Resource Allocation: The Problem of Chronic Illness in Childhood 105

EDMUND D. PELLEGRINO / Moral Choice, the Good of the Patient, and the Patient's Good 117

SALLY A. GADOW / What Good is Another Paper on The Good? No Codes and Dr. Pellegrino 139

JOHN C. MOSKOP / Allocating Resources Within Health Care: Critical Care vs. Prevention 147

GREGORY E. PENCE / Report of the President's 2003 Commission on the Fall of Medicine 163

JOSEPH MARGOLIS / Triage and Critical Care 171

ROBERT M. VEATCH / The Ethics of Critical Care in Cross-Cultural Perspective 191

ROSS KESSEL / Triage: Philosophical and Cross-Cultural Perspectives 207

STANLEY J. REISER / Critical Care in an Historical Context 215

PETER C. ENGLISH / Commentary on Stanley J. Reiser's 'Critical Care in an Historical Context' 225

NOTES ON CONTRIBUTORS 231

INDEX 232

ACKNOWLEDGEMENTS

The articles in this volume were presented at a symposium entitled "Moral Choice and Medical Crisis" at East Carolina University School of Medicine in Greenville, North Carolina, on March 17-19, 1983. We wish to express our appreciation to those sponsoring this program: the East Carolina University School of Medicine, the North Carolina Humanities Committee, the American Medical Association Education and Research Foundation, and the Arthur Vining Davis Foundations. We are grateful for the support of the faculty of the School of Medicine; the Dean, William Laupus, M.D.; the Chancellor of East Carolina University, John M. Howell; and the staff of the Eastern Area Health Education Center of North Carolina. We would also like to thank the members and staff of the North Carolina Humanities Committee and the general editors of this series, H. Tristram Engelhardt, Jr. and Stuart F. Spicker, for their kindness and help. We acknowledge our special appreciation to our secretaries, Joanne Elaine Stoddard and Margaret Gail Owens, for assisting us in preparing for the symposium and subsequent volume. Finally, we want to express our thanks to Claire Pittman for helping us with the proof-reading and editing of the manuscript and to Margaret Bunch for indexing the volume.

JOHN MOSKOP
LORETTA KOPELMAN

INTRODUCTION

The expense of critical care and emergency medicine, along with widespread expectations for good care when the need arises, pose hard moral and political problems. How should we spend our tax dollars, and who should get help? The purpose of this volume is to reflect upon our choices. The authors whose papers appear herein identify major difficulties and offer various solutions to them. Four topics are discussed throughout the volume: *First,* encounters between patients and health professionals in critical situations in general, and where scarcity makes rationing necessary; *second,* allocation and social policy, including how much to spend on preventive, chronic or critical care medicine, or for medicine in general compared to other important social projects; *third,* conflicts between or ranking of important goals and values; and *fourth,* conceptual issues affecting the choices we make. Since these topics are raised by the authors in almost every essay, we did not divide the papers into separate sections within the volume.

Warren Reich begins the volume with a parable illustrating a key problem for contemporary medicine and two very different approaches to its solution. His story begins with the "delivery" of three indigent, critically ill, foreign patients to the emergency room of a large American private hospital. Although the hospital is legally bound to care for these patients, providing long term, high cost care for them and others soon becomes a major financial strain. In response to this situation, the hospital's chief administrator proposes a "Good Samaritan" program to mobilize benefactors in support of care for indigent patients. The president of the large hospital corporation which owns the hospital, however, rejects this approach on the grounds that it conflicts with the corporation's for-profit orientation. The president announces instead a plan to use EmergiMedVan, a kind of mobile intensive care unit, to transport indigent critically ill patients to other nearby hospitals willing to admit them. As this latter plan is implemented, several hospital officials, including the administrator, resign rather than compromise their principles of beneficence and compassion in order to meet business objectives.

J. C. Moskop and L. Kopelman (eds.), Ethics and Critical Care Medicine, xi–xx.

In a commentary on his own parable, Reich highlights themes of power and the loss of a common moral tradition also found in George Orwell's *1984*. He points out that large health care corporations are assuming greater authority over both patients and physicians, and that corporate values of profit and self-preservation often conflict with traditional values of beneficence and egalitarianism. Corporations may, indeed, be able to exploit the absence of a societal moral vision to further their own interests, as illustrated by the moral absurdity of a vehicle which is specially equipped to enable hospitals to *avoid* caring for critically ill patients. Reich concludes that in order to develop a more coherent critical care policy, we must re-examine the nature and functions of our health-care institutions and find a place within such institutions for professional integrity, patient autonomy, and a balance between cost containment and equal access to care.

Tristram Engelhardt argues in another commentary that Reich has misapplied the term 'absurdity' to unfortunate situations. It is not absurd that our finite resources and ability to choose require us to pick some things over others. Whether these choices are made by societies or individuals, they limit what we can achieve. When desire outstrips capacities we have an unfortunate circumstance, but not necessarily one that is either absurd or unfair. It is our expectations that are absurd, if we hold, for example, the egalitarian belief that all can have the best of everything. We must recognize finitude and inequalities in framing a policy. Such inequalities arise from many sources, including our genetic and physical heritage and the wealth and entitlements of our families and society. These inequalities of inheritance from nature and society are compounded by our own free will. Some inequalities should not count as inequities, because they do not deprive others of their due.

Engelhardt argues that we lack one authoritative view of what constitutes the good life. It is not unfair but unfortunate when agreed-upon arrangements do not provide the best for all. Full coverage is not possible. It is not unfair if distributions respect all of their patients' entitlements, but fail to meet all their health needs. Hospitals are under no moral obligation to treat all, especially when the cost would be significant. Hence we must acknowledge their limitations in the provision of costly critical care. Engelhardt concludes that the absurdities Reich really describes are the lack of foresight, clarity, and planning in deciding how to deal with freedom and finitude.

Jay Katz, in 'Can Principles Survive in Situations of Critical Care?', assigns himself a "paradoxical task" of defending self-determination for patients in medicine by tracing its legitimate scope and limitations. It is generally understood that moral rules may be overridden in special circumstances. To apply any moral rule, including the duty to seek informed consent, knowing what constitutes an exception to it should "enhance, rather than undermine, the principle of self-determination" (p. 42). He focuses on three features of the physician/patient relationship to explore the questions of when and why patient's wishes and choices should not be honored by physicians. These are (1) the attitudes and values physicians bring to the decision-making process, (2) the attitudes and values patients bring to the decision-making process, and (3) an obligation of mutual conversation and respect in order to clarify the expectations that both physicians and patients have of one another (p. 52). He employs a distinction between the "external component of self-determination" (choice, or freedom of action) and the "internal component of self-determination" (reflection or thinking about choices). A choice on the part of either the physician or the patient may be impaired because of lack of information or skewed or irrational reflection. In addition to the well-recognized right of self-determination, there is also a duty to reflect, for patients as well as for physicians. He expresses optimism that if "the process of thinking about choices – the internal component of self-determination – is attended to with great care, the problem of a standoff between physicians and patients will be a rare event" (p. 61). Yet he favors overriding a patient's choice when two conditions are met: (1) the consequences of not doing so are grave *and* (2) the choice is seriously impaired. The more serious the consequences, the greater the patient's obligation to clarify his refusal. Indeed, physicians with their special commitments to care may legitimately insist upon understanding patients' decisions. Physicians have obligations to their patients, and they have needs that deserve respect as well. In the rare cases of disagreement aside from crisis care situations, physicians and their patients should either decide to "go their separate ways, or provide and receive care within the limits imposed by the patient" (p. 65).

Stuart Spicker argues in his commentary that Katz's underlying concern is to alleviate physician frustration and confusion over how to respond to the refusal of treatment by competent, critically ill patients. This frustration, he claims, is the result of an irreconcilable conflict between the traditional medical ethical principle to preserve and prolong life and the more recent emphasis on patient autonomy and rights.

Spicker sees Katz's strategy as an attempt to avoid this dilemma by reducing the *number* of refusals of treatment. This can be accomplished by requiring *conversation* between physicians and patients in order to overcome mutual ignorance and misunderstanding. Though he finds this an intriguing suggestion, Spicker appeals to Katz for a fuller explication of the "clinical theory" upon which such conversation will be based. This more genuine kind of physician – patient conversation will require detailed psychological explorations of both patients' and physicians' expectations.

In 'Justice and the Hippocratic Tradition of Acting for the Good of the Sick', Loretta Kopelman examines the fairness of a traditional duty of beneficence in medical ethics in general and in relation to critical care medicine. Common criticisms are that it promotes unwarranted paternalism or unjustified partiality. She argues that once this norm is understood as a prima facie and imperfect duty having an "all other things being equal" clause it can be shown that these criticisms fail. It neither entails nor does it legitimately support them. Rather, in fostering compassion and equity, the Hippocratic moral rule may be important to a just system. Partiality, when it is a result of empathetic attention to individuals, may be defensible in such a system. The strain on a just system is greatest in tragic situations when scarcity requires that some will be denied urgently needed benefits. In such cases, Kopelman argues, it is especially important for a just society to recognize the limitations of partiality and paternalism that it will tolerate. She argues that the degree of partiality, paternalism, or discretion in applying rules that is acceptable to a society varies depending upon its goals, values and priorities. Different arrangements, then, are reasonable and compatible with formal requirements of justice. Successful long-term policy, she holds, seeks impartiality, and equity. It would be unjust, however, to create unrealistic expectations based upon uninformed compassion, or to construct a system so lacking in empathy, discretion, and equity that no sensible person could use it. Kopelman criticizes several implausible policies and common responses that hinder us from forming a fair policy. Perceptions of what we want, value, think is scarce or needed, can provide, and would like to believe, must be viewed critically in framing fair policy.

In 'Clinical Ethics and Resource Allocation: The Problem of Chronic Illness in Childhood' James Perrin focuses on the difficulties of allocating health care resources for seriously ill children. He agrees with Loretta Kopelman that justice and compassion should be applied together, but illustrates in detail the tensions between them that she noted can exist. As a clinician concerned for his own patients and as a student of public policy exploring programs at the federal, state, and local levels, he recognizes the strain and difficult choices they create. Perrin notes that about 1% of all children under 16 suffer from severe chronic illness. Their care is costly: for example, an average of $150,000 yearly is spent on each of three hundred respirator-dependent children in this country. Nevertheless, care for chronically ill children has received relatively little recent attention in public policy. Perrin describes the Crippled Children's Service, one important source of funding for these children, and reviews potential criteria for distribution of services based on merit, social contribution, need, and chance. Cutbacks in services, he suggests, are generally accomplished by (a) tightening age or income criteria, (b) restricting the types of health services covered, or (c) limiting the conditions covered, e.g. covering only conditions with a good prognosis, or where a specific treatment exists, or where costs are highest. Perrin concludes that although none of the criteria he lists are very attractive, we must recognize that resources are limited and that allocation decisions cannot be avoided.

Edmund Pellegrino's 'Moral Choice, the Good of the Patient, and the Patient's Good' examines an ancient moral principle, that one ought act for the good of the patient. Though it may on rare occasions be set aside, he views this principle as "the ultimate court of appeals for the morality of medical acts" (p. 117). He examines four interpretations or meanings of the patient's good and suggests a general ranking of them. Lowest in ranking when there is conflict is the instrumental good of the patient viewed as "what medicine can achieve technically." This is called the 'biomedical' or 'techno-medical' good. Another way to view the good of the patient is the good as the patient views it. Patients who are competent are best able to say what is in their own best interest. This personal decision is contrasted with and ranked higher than the previous medical good, but lower than the interpretation of the good of the patient that centers on a capacity to choose freely and rationally. This is the good of the patient as "that which is most proper to being a human person." Finally, the fourth and highest interpretation is presented as the good of last resort. It is the conception "we hold of the ultimate good [that] is the reference standard for all decisions including clinical decisions" (p. 120). Pellegrino uses no-code orders to illustrate some of the conflicts and confusions that may occur among those failing to distinguish these different meanings. He argues that this four-fold classification and ranking will help resolve some of the current perplexities about determining the patient's good.

Three of Pellegrino's senses of patient good: medical good, personal interests, and individual choice are examined by Sally Gadow in her commentary. Gadow acknowledges that distinguishing these three meanings of good can counteract a tendency of some physicians to focus exclusively on medical goods. She suggests, however, that the three may ultimately reduce to autonomy, since Pellegrino clearly subordinates medical good to patients' interests, and he describes patients' interests as anything patients freely choose. If autonomy is the primary patient good, however, it is one that can be very difficult to attain in some cases. Thus, Gadow returns to Pellegrino's notion of medical good, narrowly understood as cure or palliation. For Pellegrino, CPR (cardiopulmonary resuscitation) represents a medical good only when correcting the immediate crisis (cessation of vital functions) contributes indirectly to improvement in pre-existing pathology, that is, when the patient's condition is not hopeless. But, Gadow points out, the judgment that a condition is hopeless, or not worth prolonging, is one that can only be made by competent individuals for their own lives; medical good must, therefore, still be subordinate to patient autonomy. Finally, she proposes that CPR be understood as *rescue* or as focusing only on alleviating the immediate crisis. This approach makes CPR "the most ordinary of human responses" (p. 143) – automatic, unquestioned, and morally obligatory in the absence of indisputable evidence that persons do not wish to be rescued.

In 'Allocating Resources Within Health Care: Critical Care vs. Prevention', John Moskop examines arguments for and against a high level of support for critical care medicine. Moskop criticizes arguments favoring critical care based on a special symbolic value of critical care, on the physician's commitment to individual patient welfare, and on the urgency of critical care needs. He notes that a number of recent writers have called for a shift of resources from critical care to other areas, such as preventive medicine, self-care, and basic primary care, on the grounds that these areas will make a greater overall contribution to health. Moskop argues, however, that an accurate estimate of the cost-effectiveness of critical care awaits further research, and that what data is available does not present a clear picture either for or against continued support. He acknowledges the major contribution of basic preventive measures to improving health in the last three centuries, but questions how much further benefit can be expected from prevention in developed countries in the future. In these countries today, many diseases are associated with choices and habits, such as smoking, drinking, overeating, and physical inactivity. Because individuals are often unwilling to give up these things, and society is reluctant to prohibit them, we should expect a continuing high incidence of serious illness requiring critical care. Moskop concludes that the arguments for increasing support for preventive medicine are no more successful than those favoring critical care.

Speaking through Gregory Pence, Stewart Lawrence Randolph III – a visitor from the year 2003 – offers a glimpse at the future of health care in America. He argues that several trends already evident in 1983, including increases in Social Security, VA and AFDC payments for medical services will, if continued, place serious strain on taxpayers' ability to support the health care system. Significant changes in our methods of providing and financing health care may, however, be blocked by a fundamental moral impasse between supporters of welfare-oriented and libertarian social policies. The result, Randolph or Pence warns, could be bankruptcy and breakdown of the American health care system – the Great Crash of Medicine in 2003. Such a breakdown, he concludes, may yet inspire a new vision of scientific medicine and a return to the ancient virtue of charity.

Joseph Margolis begins 'Triage and Critical Care' by distinguishing between medical triage and market triage. Understanding market triage as the sorting of things into good, bad, and fair without any implication of scarcity and medical triage as the allocation of scarce but important medical services, there is a striking difference between them. The initial problem of medical triage is a *tragic choice*, since some of those who could be helped will not be helped. Margolis discusses various solutions to this problem of medical triage offered by advocates of both patterned and nonpatterned moral principles. While patterned approaches such as egalitarianism, utilitarianism and contractarianism are currently more popular, they do not, he argues, eliminate the evil of the initial medical triage situation. Non-patterned approaches, which acknowledge the inevitability of tragedy or evil, are, Margolis argues, no weaker. While a particular solution, patterned or not, may be quite justified in some cases, there is no way to derive a general solution to medical triage by direct appeal to moral principles or human nature or natural human rights. Rather, whether the advocate of triage favors the patterned principle that dictates a particular solution under the given circumstances (i.e., a verdict) or a non-patterned principle, we are at a stalemate in choosing between the various solutions unless we look at time and place. For both kinds of solution, though quite different, "are honorable and impressively convincing to different populations in different circumstances . . . . If this means there are no fixed, substantive, determinate moral laws, so be it. Or if the moral law is generous enough to tolerate such divergencies, then it cannot be very far from being completely vacuous" (p. 186).

Robert Veatch, in 'The Ethics of Critical Care in Cross-Cultural Perspective', argues that the dominant ethical questions about critical care medicine cluster around two kinds of problems. The first concerns issues of autonomy and paternalism where a common question is: "How can the rational individual express his autonomy by deciding to escape the assaults of aggressive but agonizing tinkering by medical professionals desperately trying to overpower nature when the patient would prefer to step back and let nature take its course?" (p. 191). The other cluster of issues concerns the fair allocation of critical care and is characterized by questions such as: "How can each member of the community get his or her fair share of the wonders of medical science?" (p. 191) Veatch asserts that while there is little doubt these questions are the dominant issues in the United States, there is also little doubt that they do not dominate medical ethics discourse in other countries as they do in the United States. In trying to account for this, Veatch discusses three conditions: high resources, technological skepticism and liberal individualism. He examines these conditions in the health care systems of Sweden, Poland, Cuba and the United States. He argues that the United States, more than any other country, embodies all three conditions and that the combination of them is what makes justice and autonomy substantial moral issues in the delivery of critical care.

Commenting on Margolis and Veatch, Ross Kessel argues that the issues they identify in the allocation of scarce and critical medical services cannot be separated from more general issues about the just distribution of health care. He interprets Margolis's view to mean first, that some kinds of sorting, but not others, necessitate neglect of some of the needy; second, that "the neglect of some of the needy cannot be morally justified and that responsibility for this neglect must be accepted" (p. 210). Kessel argues that the "lack of adequate justification for triage decisions in medicine will only deepen if we extend the discussion to include ordinary as well as extraordinary measures of health care" (p. 207). He regards as problematic Margolis's attempt to limit the discussion of triage to extraordinary care. When we are discussing grave consequences and where need outstrips the expected resource, we have no clear line between the extraordinary and the ordinary.

Kessel agrees that Veatch has identified two important issues dominating the current medical ethics debate in the United States and that the conflicts over issues of autonomy and justice, or freedom of choice and access to health care services, do seem to be unusual in other countries. As he did with Margolis, Kessel holds that these reflect general issues of providing health care in a society, not simply critical care. He argues that a greater acceptance of limitations or services, of paternalism and of support for a social responsibility for the needy in other countries may be rooted in a greater homogeneity of the people. The valuing of autonomy within medical care in the United States may be the result of our pluralism. The differences noted by Veatch, then, are part of larger differences that make our society unique.

Stanley J. Reiser's 'Critical Care in an Historical Perspective' examines how physicians in three different periods responded to providing acute medical care. In the Hippocratic era of Greek medicine, the goals were to lessen suffering and to use therapies to assist nature's restorative powers. Therapies were regarded as having a circumscribed scope and were typically non-invasive, such as diet, exercise, or bleeding in small amounts. Since the principal healer was Nature, excessive intervention was considered imprudent. If the Greek physicians sought to be nature's assistants, the American physicians of the first half of the nineteenth century wanted to conquer it. In contrast to the restraint and moderation of the ancient Greek physician, these American physicians practiced "heroic therapy," including blistering, copious bleeding and agents to induce vomiting and purging. In the twentieth century, growth of technological capacity to sustain critically ill patients has, of course, lead to the debate about when such interventions are appropriate. This debate includes moral and legal issues concerning the proper limitation of the use of technology out of respect for the rights and dignity of the patient, and troubling concerns about the allocation of scarce medical resources. Reiser argues that in trying to resolve these issues "we can be helped by ideas proposed over two millenia ago by the Hippocratic Greek physician" (pp. 222–223). Their concern for balance, for an awareness of their own limitations, and for providing a rational account of the goal of therapies could lead us to a synthesis of our technologic and humanistic concern.

In his commentary on Reiser, Peter English points out that there has been very little historical analysis of the topic of critical care. History can contribute to this topic, he suggests, by investigating three crucial elements of the meaning of the word 'critical' in medical contexts: (1) the need for judgment based on skill or knowledge, (2) the importance of timing, and (3) the presence of uncertainty and risk. Greek medicine offers a good starting point for such an investigation, since timing was a major concern of ancient Greek physicians. English disagrees with Reiser's claim that the Greeks avoided critically ill patients, holding that they distinguished between critical and terminal illness. He suggests that despite Reiser's jump from the ancient Greeks to the 19th century, critical care issues have always been present and that the treatment of plague and wound surgery are topics in the intervening centuries especially deserving of careful study. Finally, English cautions that 19th century therapy was not monolithic and offers the physiological surgery movement as an example of an ethically motivated response to the excesses of late 19th century surgical radicalism.

Nor is the response of the twentieth century monolithic, as our authors show. We hope their views will help us confront and solve the moral and social problems posed by critical care.

JOHN C. MOSKOP
LORETTA KOPELMAN

WARREN THOMAS REICH

A MOVABLE MEDICAL CRISIS

On January 14, 1983, a small aircraft landed in an old World War II naval airfield sixty miles from the Atlantic coastline in a southeastern state. Four passengers deplaned. One of them, a healthy young man in his thirties, walked briskly past the only three private planes parked in the field, entered a telephone booth and called a taxi. The escort assisted his three very sick companions – two men and one woman – into the cab and asked the driver in broken English to rush to the nearby Pilt Hospital Emergency Room.

On arrival at the hospital, the taxi driver nervously assisted the rapidly failing trio into the emergency room before departing. The healthy escort situated his three compatriots in a slumped but comfortable position in the emergency waiting room, rang the service bell, and departed. He called another cab, raced to the airport, and took off in the waiting plane, which flew out over the Atlantic before heading south.

Six weeks later, on February 26, Doctor L. Jonathan Harrison was presiding at a staff meeting of the Pilt Hospital Concentrated Care Center, at which the cases of the three patients were being reviewed. Dr. Ralph Schleiker, medical director of emergency medicine, explained that on their arrival in the emergency room the patients had been in shock when discovered by the triage nurse. In the work-up, they had to be dialyzed for purposes of diagnosis. All three had end-stage renal disease, and now all three had been in Pilt's intensive care unit for six weeks.

The staff was under great pressure from several sources. First, the hospital administration insisted they could not absorb any more of the aliens' bills, now totalling $185,000. None of the three patients spoke English; none had any money or insurance. The hospital's financial officer said: "There's a limit to the amount of debt we can absorb. Ultimately, we pass those expenses on to the customers, but we have already gone beyond the limits of what we can handle by our usual robinhooding methods. Not that we steal from the rich to give to the poor, as Robin Hood did; we pass on the expenses of indigent persons to the unconsenting paying public. But we can't robinhood for these Salvadorans any longer."

J. C. Moskop and L. Kopelman (eds.), Ethics and Critical Care Medicine, 1–10.

A second source of pressure arose from the fact that Pilt Hospital had only five adult dialysis machines. When the sick Salvadorans were dropped at the emergency room, space for them on the machines happened to be available; but normally the five machines are kept busy at all times with local referrals. The purpose of the Concentrated Care staff conference was to make recommendations on removing some patients to another facility or to outpatient care to make room for three kidney patients from the local county – two patients fully covered by Medicare and private insurance, and one uninsured patient whose medical and hospital fees had been paid in advance.

Third, the medical staff was quite disturbed that without medical records for the patients they were losing a valuable opportunity to continue their investigation into treatments for renal disease. The hospital administration had been able to discover only the names of the patients and the fact that they were illegal immigrants from El Salvador. The committee resisted the pressures; the Salvadorans were not denied care.

At a meeting on March 16, the hospital administrator, Dr. Hilaire Weeks, formerly an internist at the Pilt Hospital, confessed to the hospital legal counsel, Mr. J. Michael Fox, that she was exhausted by the Salvadorans' case. "Personally," she said, "I don't see why we had to accept them as patients. If I had been on call in the emergency room at the time of their arrival, I would have been tempted to turn them away." Fox responded, "You have no choice. There is a legally sanctioned expectation that you accept all patients in extremis who present themselves at an emergency room. If you seriously jeopardize their health by turning them away you can be sued – perhaps on grounds of negligence, perhaps on grounds of discrimination."

Weeks responded: "I understand this is not that clear legally; maybe we could take our chances on both counts. After a certain point, the risk of liability for negligence is just a risk. The combined bills of these three patients is now $275,000. We may be able to discharge them to outpatient care within the next week, thus reducing but not removing the expense. I'd almost be willing to take my chances on the discrimination charge. What the hell, they're illegal aliens. Don't our obligations to Americans come first?" Mr. Fox added: "I've been able to find out more about the Salvadorans. Three nurses and two physicians in El Salvador financed the patients' trip by private plane to that old airstrip out on County Line Road, with strict instructions to go directly to our emergency room where they knew we could not turn them away. The confederates back home in El Salvador clearly had a humanitarian motive. What I have not been able to verify is the report that they also had a political motive. The five health professionals who foisted the deal are all insurgents who are deeply resentful of Washington's policy of sending armed helicopters for national troops instead of sending medical resources to the peasants. But I'm still working on that aspect of the case; I'm in daily contact with the legal division of our national office."

Dr. Weeks, the hospital administrator, retorted: "But back to my point. We don't strictly owe anything to the aliens. They are not our fellow citizens and they haven't bought into our plan." "I'm not so sure about your ethical reasoning," attorney Fox responded. "The International Civil Liberties Union has already heard of the case and is watching the level of treatment received by our Salvadoran patients. Seems they are looking for a case where they can argue for constitutionally-based equality of medical treatment for all residents, whether citizens or non-citizens, whether legal or illegal aliens. The whole case is just as absurd as it can be. But don't worry about it, Pilt is personally keeping an eye on the case. We are in daily contact by phone."

Four days later, Dr. Weeks, meeting with her finance officer, complained about a new expense: "The patient that we accepted from the Jefferson County Community Hospital seventy five miles down the road – Belinda Taylor – is now causing us a lot of problems. She is costing us $935 per day and will almost certainly die within fourteen days, though she probably would have died in about two days if she had stayed at Jefferson. They shipped her here because we have the reputation of being such a marvel of high technology with a huge budget. So *we* are paying about $11,000 to keep her alive an extra twelve days. The word has gotten out around town that her family shipped her here because they ran out of money to take care of her down in the country. The local citizens are up in arms; they regard this hospital as the hospital for *this county* and deeply resent out-of-county folks coming in here to use their sophisticated medical resources. I'm having a worse public relations problem with this Jefferson County patient than I am with all three Salvadorans put together.

"The Belinda Taylors of this world are bleeding our system dry," Weeks continued, "and to what good purpose? Last month federal Medicaid support was cut by 21%. Pilt Hospital had to absorb 40% of the entire state reduction because we are private and the centerpiece of the best private medical school in the southeast. All the remaining forty-one hospitals in the state absorbed the remaining 60% of the reduction. Our nearby Wayte Memorial Hospital has full Medicaid coverage through state Medicaid. They should be getting all the Belinda Taylors of the state."

Dr. Weeks continued to press her case with her finance officer: "While I don't think we have any strict obligation to treat aliens, their case has made me think more about our obligations to strangers who are Americans. It bothers me as a doctor that there are strangers in our midst – fellow citizens who are deprived of critical-care medicine because they can't afford it. What good is a hospital if it is incapable of occasionally being a Good Samaritan with critical-care medicine? I propose that we begin with local associations, organizations, and individual benefactors to discover our local potential for Good Samaritanism in developing reserves of cash and services for the needy person who falls between the cracks of payment schemes. Then we should go to the state and regional level, and then take that foundation and go to the Pilt organization and see what concessions they would make."

Pleading desperately with his hospital administrator for more funds just two weeks later was Dr. Aaron Davidson, director of the Pilt Hospital Neonatal Intensive Care Service. "We just don't have enough beds for babies needing intensive care," he said. "We have to ship them all around this part of the southeast, to five different cities, one of them over 150 miles away. The federal government has put us in a bind that is totally absurd: On the one hand they cut funding for neonatal intensive care by 40%, and then they threaten to cut off all federal funds from any hospital that lets a handicapped infant die. Because of the government's action, we cannot save the salvageable babies that should be saved; but we are legally threatened if we let non-salvageable babies die. Sometimes I don't know what our society expects of us. I'm concerned about saving the babies that *can* be saved, and that's why I'd like to be part of your effort to raise private Good Samaritan funds to fill this need."

Mounting problems at the Pilt Hospital had not escaped the attention of higher authorities. On September 27, J. Bimsley Pilt, President and Chief Executive Officer of the J. B. Pilt Company Health Services Corporation, was addressing her executive staff: "I have two things to say, first regarding some general corporation policies, and second, about a problem we have in one of our hospitals.

"First, to review our current major policies. Our primary objective is the preservation and prospering of the corporation. We owe that to our investors, as well as to our consumers who want the best possible medical care. Our rapid growth is unequaled in the business of owning and managing acute-care hospitals. With about 87,450 beds in 592 hospitals, revenues from our operations exceeded $5 billion in 1983. Our company's earnings per share are currently doubling every three years, a rate that I propose to maintain.

"We are expanding and diversifying in some interesting ways. We now have forty-eight projects in eleven foreign countries. Within five years we will dominate Western Europe's private hospital market, catering to those Europeans who are dissatisfied with various national health plans. Acute care," Pilt continued, "is the leading edge of the wedge of our corporation, but other segments of the health-care field follow along. Last December we purchased a 24% ownership of a chain of for-profit health maintenance organizations, increasing the number of our employees to 141,000. I want the term 'Piltco Healthcare' to become a household word by the end of the decade."

Pilt continued: "As an accountant-turned-business-executive, I have tried to establish some clear, basic principles of operation. We have three rules in this corporation. First, we will not compromise our high standards of quality medical care. If a hospital doesn't have the potential to meet the quality standards we demand, Piltco will not buy out the hospital; and if the hospital doesn't continue to measure up to our quality review criteria, we will let it go. Second, the consumer must share in whatever is the normal cost of health care. Health care costs whatever it really costs, and insurance companies should not be the only ones carrying the risk of that cost; consumers should bear that risk as well. Third, every decision must be made from the standpoint of what its long-term impact will be on the corporation.

"That third rule leads me to the second item of business, the hospital in Carolina that has been having so much trouble and notoriety. That's the hospital that is treating the Salvadorans. I am thoroughly acquainted with the attitudes and actions of all the key personnel involved in that hospital.

"Dr. Weeks is the hospital administrator. I do not fault her for the way she has handled the Salvadoran cases. That was forced on her; I call it an act of international medical terrorism. In a very real sense they're holding one of our institutions hostage. But considering the enormous financial pressure that has been placed on her hospital by those aliens, she is absorbing too many costly but hopeless cases from other parts of the state. In response to some of these problems she is also making some mistakes. She has declared a policy of charity that is not acceptable to the corporation. She announced the launching of a 'Good Samaritan' Program for the assistance of indigent people from the state who need expensive, critical-care medicine. One thing I have learned over the years is that Texans know a helluva lot more about the Bible than do Carolinians. Someone should tell Weeks that the parable of the Good Samaritan is about one ordinary citizen being a good neighbor to another citizen. What kind of 'medical care' did the Samaritan give to the Jew? He rubbed oil in his wounds. And I want you to know that I am willing to rub oil on the sore ass of every Carolinian before I will accept any so-called 'Good Samaritan' policy in any of our acute-care hospitals. We should not be giving the impression that doctors give away a hospital's expensive, acute-care services.

"I don't mean that there are hospital patients who won't get treated because they can't pay. The indigent are treated in our hospitals. If the patient cannot pay and public reimbursement is not possible, we shift the cost of indigent care to the other patients who are able to pay their own bills directly or through third-party coverage. If the local hospital cannot carry the burden, the corporation can shift the expense to paying patients in other, more prosperous sections of the country. Thus, the care of indigents is carried out entirely within a business framework, not in a charitable framework. We have only one basic policy, and that is a for-profit policy. I believe we must be very careful about which aspects of our policies are publicized. I don't mind letting the world know that we treat a lot of non-paying patients, but I wouldn't want too much sunshine on the ways we shift payment among patients. Furthermore, in my opinion it would be detrimental to the corporation's well-being to allow any of its hospitals to have a substantial, local-charity policy. I am telling Dr. Weeks that she cannot go ahead with her Good Samaritan plan.

"Finally," Pilt continued, "I have decided on an innovative solution to several of the problems that are plaguing our Carolina hospital. If the solution works well there, we may try it in some other parts of the country. Basically, the problem is caused by the obligation to accept patients in emergency rooms; by the sharp increase in the number of unannounced patients seeking access to high-technology acute care through an emergency room; and by the sharp decline in revenues available from federal sources and from the patients themselves where unemployment is high, coupled with state-level discrimination against private hospitals. Actually, the legal pressure is not to accept all patients in emergency rooms, but as I understand it, basically only those patients who are in extremis and who cannot be safely transported to another place for medical treatment.

"In my new policy, we will turn no patients away from the Pilt Hospital emergency room. When the hospital's capacity to handle critical-care patients is strained, some of those patients will be placed aboard a van that will be called EmergiMedVan. The first EmergiMedVan is now being equipped at a cost of $950,000. On board EmergiMedVan will be sophisticated life-support systems and monitoring systems that will be in direct contact via radio and telephone with the hospital's intensive care personnel. In a word, EmergiMedVan's patients will have high-quality medical care.

"EmergiMedVan will transport three patients at a time, as needed," Pilt explained, "to the nearest hospitals where they can be transferred, beginning with the more favorably funded, nearby public hospital. There are forty-one other hospitals in the state. We hope that each patient will find satisfactory treatment within, say, the first two or three hospital stops at the most. But if those hospitals refuse to accept them, EmergiMedVan will be fully equipped and prepared to carry the patients indefinitely through a large circuit of hospitals. If necessary, the patients could be returned to Pilt, but I expect we'll be able to find a new facility for each of them.

"My solution has many merits," Pilt commented, "not the least of which is that it permits us to fulfill all our legal and ethical obligations without jeopardizing our humanitarian posture, or losing our high-technology image.

"EmergiMedVan will serve another constructive purpose. The Pilt Company wants to focus on those hospitals that take care of the more intensively ill patients. EmergiMedVan is a means of maintaining the hospital's reputation of being a first-rate, tertiary-care hospital during a transitional period. Confidentially, I view 'Supervan,' as I like to call it, as part of an economic mechanism that will help carry us over to a new realistic era in hospital care, an era in which some lower level of standard medical care will be given in public hospitals, while the more sophisticated and expensive care will be available in the type of private hospitals that the Pilt Company is purchasing.

"Finally, EmergiMedVan is fully endorsed by our chief legal counsel, who believes that the Justice Department and the State Department will be pleased with a hospital policy whereby we will not have to turn away desperately ill foreign insurgents at a time that would be most embarrassing to the U.S.

"One difficulty has been clarifying the legal issues about consent. There will be special informed consent procedures in EmergiMedVan," Pilt continued. "We have anticipated the possibility that a patient's treatment modality may have to change in transit. Because of the limited size of 'Supervan' it will not be possible to consult carefully with the patients on board nor with their next of kin. But for some time now, that particular hospital has been doing a 'values history' on all patients; and those values histories have now been computerized. When treatment alteration is indicated in EmergiMedVan, we will consult the computer. If the patient's values history indicates that, given prospect type X, the patient would be complacent, we will assume consent. If, on the other hand, the history shows the patient to be a fighter in that regard, we might rightly assume refusal. This consent procedure has been thoroughly reviewed and approved by the company's Ethics Committee here in Houston. The Committee has determined that consultation with our recorded histories is ethically superior to consultation with living relatives on behalf of an *incompetent* patient, inasmuch as our method makes possible a consultation with the patient's *own* set of values, not those of the relatives, and thus more perfectly respects the patient's autonomy. And although it is true that actual consultation with a *competent* patient on board EmergiMedVan would, all things considered, be preferable to consultation with his or her recorded values, the Ethics Committee has rendered its opinion that actual consultation with the patient is, in these unusual circumstances, supererogatory in view of the need to reduce costs and conserve valuable life-saving space in 'SuperVan'."

News of the company's new EmergiMedVan policy spread quickly up and down the halls of Carolina's Pilt Hospital. And reactions were swift in coming. On October 15, the hospital's ombudsman - Professor Harold Mill, who oversees patients' interests - resigned. He made this statement to the hospital ethics committee: "I must resign because of my commitment to the patient's autonomy - not because informed consent rules are being ignored in EmergiMedVan but because the unique sentiments of the patient and family will not be heard. On the other hand, I can't stay and fight the issue, because I am convinced that EmergiMedVan is a sensible solution to the management problems of this hospital."

A few days later, at a meeting of the hospital's neonatology staff, Dr.

Aaron Davidson, Chief of Intensive Care Nursery, resigned. He gave this reason: "I have always suffered with my patients and cried at the death of any infant who died in my unit. Long ago I promised myself that I would resign if I ever quit crying, but I always assumed that if my tears dried up it would be due to burn out. Now I must resign because I have been deprived of the possibility of being burned out. How can I sustain compassion for patients, not knowing whether they will return from an intensive-care orbit? As a physician, I cannot function in a system that controls the very intuitions that tell me when compassion is appropriate."

On the last day of October, Dr. Weeks, the hospital administrator, confided in her husband: "I don't mind having my ideas rejected by the corporation, but I do feel compelled to resign as hospital administrator for two reasons: First, the hospital has clearly lost its autonomy: How can it function responsibly when so many important decisions affecting it are made elsewhere? Second, I have always felt there was a strong compatibility between being a physician and a hospital administrator. But when my idea of trying to samaritanize the hospital and the local community was so vigorously rejected, I became convinced that there just isn't room any longer for my own feelings about combining traditional professional beneficence with the hospital business. Maybe others can make the combination more successfully than I can."

On January 3, 1984, a brilliantly white EmergiMedVan hummed as it pulled away from its dock adjacent to the Critical Care Center and rolled down its long ramp in a quiet launch into its first orbit of mercy.

That evening on the local Greenburg television news program, the announcer read: "Pilt Hospital today upgraded its medical intensive care services with EmeriMedVan, the country's first fully equipped intensive care van. EmergiMedVan will greatly enhance patients' safety and health during necessary transport."

Georgetown University School of Medicine
Washington, D.C.

NOTE

[1] This paper was prepared while the author was a Fellow (1982–83) of the National Humanities Center at Research Triangle Park, North Carolina. The author expresses his gratitude to the National Humanities Center, Georgetown University, and the Hillsdale Fund for their support of the project, and to colleagues who made helpful comments on an earlier draft of this paper: Houston Baker, Larry Churchill, Harmon Smith, Lance

Stell, Joanne Trautmann, and especially A. Kenneth Pye, whose observations provided the stimulus for the development of this writing and the commentary constituting the next chapter. Nonetheless, I take responsibility for what is written here.

WARREN THOMAS REICH

MORAL ABSURDITIES IN CRITICAL CARE MEDICINE: COMMENTARY ON A PARABLE

The foregoing tale of EmergiMedVan may serve as a parable for the burgeoning moral world of critical-care medicine. While that narrative embodies the main points that I want to express, I would like to add a postscript that suggests more explicitly the themes and agenda for moral choices in critical-care medicine.

As this paper's title suggests, the parable of EmergiMedVan portrays moral absurdities in critical-care medicine. Many aspects of critical-care medicine are rightly called absurd because they are incongruous, or "ridiculously incongruous," as the dictionary defines the term; and others are absurd because they are irrational.

The absurdity in contemporary critical-care medicine is implied in the over-arching theme of the parable, for it appears to me incongruous that a rational society should simply discover that a highly valued and personal area of life (health-life-death) is increasingly controlled by institutions that are rapidly proliferating and expanding their power over us. Furthermore, it is ironic that this discovery should be made at a time when we are becoming more deeply aware of the disappearance of those traditions that have given us a shared assurance of knowing right and wrong.

These two disjointed themes of power and of the loss of a common moral tradition are found in another narrative, George Orwell's *1984* [24]. Thus, the parallels between the subject matter of the parable and that of Orwell's work are much more profound than a mere coincidence of dates. Both these themes are found in the larger world of American health care but are exacerbated in high-cost, critical-care medicine. We face two themes, then, that are very similar to those that were bothersome to Orwell.

(1) *Power.* The first of these common themes is that of increasing power and shifting autonomies. Among those autonomies the first is the autonomy of the *medical profession,* which has acquired a privileged status in American society[2] - a monopoly achieved through a variety of economic and political strategies [3]. Paul Starr has described the

J. C. Moskop and L. Kopelman (eds.), Ethics and Critical Care Medicine, 11–21.

"sovereignty" of the medical profession in terms of its scientifically-based authority over patients and its monopoly over the structure and finances of the health-care system, especially through its control over the relation of patients to hospitals, pharmaceutical companies, and use of third-party payment [32]. *Insurers* became an important power base in medicine, but generally allowed doctors to retain the autonomy of independent entrepreneurs, passing on the costs of professional autonomy to the consumer.

However, the scope of the autonomy of many practicing physicians was eroded by the increasing autonomy of the various hierarchies of *medical specialties,* which began to control both supply and demand in critical-care medicine. As they acquired the role of transmitters of scientific advances in medical treatment, they were able to shape the demands of the ill; through certification, restricted hospital privileges, and in-hospital rules they also came to control the supply.

In addition, the autonomy and authority of *the hospital* have grown as it became the dominant health-care institution. With liberal government support, there has been a tremendous growth not only in proprietary hospitals, but also in *other proprietary health-care institutions:* nursing homes, dialysis clinics, home-care health businesses, and emergicenters. The institutional autonomy of hospitals has generally protected the professional autonomy of physicians; hospitals themselves, however, (and health maintenance organizations) are becoming more heteronomous, gradually sacrificing their authority to expanding *health-care corporations* ([32], p. 477). These new institutions - part of what Relman calls the "new medical-industrial complex" - raise the question whether medical care will be provided primarily in the interest of the stockholder or that of the patient and the public [29]. Certainly, in the corporate-owned, multi-institutional, for-profit hospital chains, individual physicians are deprived of much of the influence they have wielded over hospital policy ([32], p. 447). Indeed, there are indications that physicians practicing under corporate management will experience a more profound loss of autonomy: for example, "the corporation is likely to require some standard of performance, whether measured in revenues generated or patients treated per hour" ([32], p. 446).

In addition to this array of institutions that function as enterpreneurs competing for power, medicine has been strongly influenced by the "rationalizers" - institutions whose function is to regulate the structures, quality, and cost of health care. These include the federal government, federally-sponsored organizations, states, businesses, business-sponsored organizations, employers, etc. Authority and autonomy are shifting rapidly among these *regulatory agents*. The role of the federal government in financing (e.g., through Medicaid and Medicare) and regulating (e.g., through Professional Standard Review Organizations) health care is well known. Less well known are the regulatory mechanisms now increasing in the private sector, especially the role of health-care corporations as regulators. In that context, it is likely that clinical performance of physicians will be assessed by statisticians who utilize quality-control programs in corporation headquarters ([32], p. 447). Thus, increasingly, medical decisions will become, or at least become strongly influenced by, business decisions. The rise of a corporate marketing ethos in medical care is now leading to designs for corporate techniques for modifying the behavior of physicians, "getting them to accept management's outlook and integrate it into their everyday work" ([32], p. 448). This development will produce a subtle but profound loss of autonomy for the medical profession - not only over the social and economic dimensions of medicine, but also over the technical, medical content of their practice and the very ideals of professionalism ([32], pp. 447-448; 495, n. 72).

Furthermore, the institutions that serve as rationalizers of the system not only regulate American health care but also use their power to reorient the system. It has been claimed, for example, that the Rockefeller Foundation had an enormous influence over the medical profession in America through its funding activities that were undertaken with the partial but pronounced intention of molding a medical profession that would promote corporate influence over American institutions [1]. Similarly, the ascendancy of a market ethos in American medical care can be expected to subordinate health care and its distribution to considerations of marketing and profit more strongly than before and without the cushion of the professional ideals of medicine ([32], p. 448).

In the midst of these shifting economic, social, and political powers are individual persons, eager to exercise their own autonomy. People want control over their own health affairs and to assist their relatives and friends in the personal control of their lives. Yet *the individual sick person* may be quite helpless, because of his exclusion caused by poverty, or simply his powerlessness caused by having too little control over the system. Similarly *individual professionals* – like the parable's Dr. Weeks, the hospital administrator, and Dr. Davidson, chief of neonatal intensive care - seek an autonomous acting-space where they can function as true professionals with moral integrity. Yet those very professional endeavors are rendered dubiously effective by the shifting autonomies of institutional health-care powers.

That, then, is the first theme that modern medicine shares with Orwell's world, for it is a world of controlling powers and voracious autonomies that creates a sense of moral confusion. A sense of absurdity is heightened by the fact that corporate health-care forces have arisen in an unheralded way, without prior study or reflection on the part of the public as to how these moves would affect moral agencies and ethical values in medicine.

(2) *Loss of a common moral vision.* The second theme, common to critical-care medicine and Orwell's *1984,* is the disappearance of our shared belief in a firm basis for knowing rights and wrongs [17]. The era of moral incertitude is, of course, not of recent origin. In the 16th century Montaigne wrote:[3]

> What am I to make of a virtue that I saw in credit yesterday, that will be discredited tomorrow, and that becomes a crime on the other side of the river? What of a truth that is bounded by these mountains and is falsehood to the world that lives beyond? ([19], p. 437)

As the parable of EmergiMedVan may have illustrated, our society is characterized by pervasive moral ambivalences, which is partly due to our not being clear about what sort of society we want to be.[4] The incongruities and conflicts among our principles bear out this analysis. For example, as will be noted again below, the egalitarian principle, which has strong legal standing in the emergency departments of our hospitals [22] is deeply at odds with the autonomy-based theory of rights that would protect individual holdings from societal redistribution [23], and with assumptions about the autonomy of the political community whereby it can determine as it will what services it will render to whom it will [2].

If I am correct in observing that we are now experiencing the juxtaposition of these two themes - powers that render us in some important respects powerless and the lack of a clear sense of rights and wrongs for coping with them - then we do indeed face some of the same agitating questions as those posed by Orwell [31]. Transposed to the scene with which we are concerned, those questions are: In the economic, political, and institutional world of critical-care medicine, do we commit ourselves to action without a clear moral vision? Do we commit ourselves to a set of moral beliefs without the power to act? Does the sheer complexity of a new medical era characterized by the power bases of critical-care medicine lead one to succumb to moral exhaustion? EmergiMedVan becomes the symbol of the manipulation of existing health care structures for the perpetuation of power – a manipulation that is effective because it exploits the very absence of societal moral vision.

The over-arching concern of the parable, then, is the moral shape of our institutions and the moral vision that molds them. If we view ourselves as individuals who seek a coherent and rational explanation of things, we are struck with a sense of absurdity. And we are justified in referring to the absurdities of critical-care medicine, for the moral dimensions of the medical institutions that are being shaped by and for critical care are irrational and incongruous - in some respects ridiculously incongruous.

There are some who would want to draw conclusions about the inevitable moral skepticism or pessimism that can arise from a perception of absurdity in our social institutions. Pessimistic conclusions of this sort have been drawn form Orwell's writings [31]. I would not want to draw those conclusions, because I think we should be about the business of re-examining the purposes and functions of our institutions, including those that are only now taking shape, and giving a rational and moral direction to them. But in proceeding to examine critically the individual quandaries and principles related to critical-care medicine, we should continue to be aware of the incongruities in the individual concepts and principles with which we are dealing. Let us consider these more explicitly.

Corporate values and responsibilities. An initial problem is how to reconcile the ethics of corporations and other large institutions with the ethics of medicine. We need to consider whether there is a discrepancy between a hospital corporation ethic whose highest value is its own preservation and financial prosperity and the traditional ethic of the medical profession that pledges a deep commitment to serving the sick and injured without discrimination [15, 28]. What voice should be given to the public (i.e., those potentially needing the care) and the members of the medical and other professions in shaping or changing the shape of health-care institutions and their policies?

Some hospitals have demonstrated the ability of an institution to make major allocational decisions, restricting programs in critical-care medicine partly on the basis of discriminations among the benefits that a hospital should be giving to the public [16].

At the theoretical level, important efforts have been made to examine collective responsibility in medicine and health care, i.e., moral responsibility in contexts that bind together physicians and other professionals with collectivities such as the hospital, the health-care team, the medical school, and organized professional societies [20]. Yet while it is important to reflect on the nature and application of concepts of collective responsibility in already-existing institutions, it seems less than reasonable simply to witness uncritically the formation of new health-care institutions - hospitals, corporations, associations, etc. - without asking the question: Where *should* the *collective* locus of health-care responsibility lie? The question becomes increasingly urgent as one notes that health-care institutions arise or take new corporate shape on the impetus of such diverse forces as critical-care technology in search of use, investment potential in search of profit, and the desire to serve the most important health needs of the public. The attempt to develop the desirable shape of our health-care institutions would seem to require articulation of theories that would best undergird an ethic of corporate responsibilities, and ultimately reflection on what sort of society we want to have.

Professional integrity. There is a kind of absurdity in the actions of a director of an intensive-care nursery and of a hospital administrator who feel compelled to abandon their positions so as to preserve their own personal integrity, but at the expense of the integrity of the institutions in which they are immersed. Thus, the problem of integrity to which this parable draws attention is not simply that of the *difficulty* of being true *both* to one's own moral self (together with its presuppositions in certain models of professional ethics) *and* to the expectations of a health-care institution [18], or even of reconciling conflicting responsibilities at personal and institutional levels [4]. Certainly, those moral problems are important for any discussion of the ethics of institutionalized critical-care medicine. But the problem of integrity that I would want to highlight is found where the institutional structures of medicine render *irrelevant* the moral character of the individual professional and the endeavor to ground one's character in an autonomous commitment to professional values.

Autonomy of the patient. The informed consent policy on Emergi-MedVan is suggestive of a genuine and pervasive problem. It is no longer adequate to discuss self-determination of patients merely in terms of reduced autonomy in hospitals, for what is at issue is personal autonomy within the broad *societal* structures of medicine and within the narrower institution of *hospitalization* (not within the hospital as institution, but within the institution of hospitalization).

At the societal level, we would do well to consider the relevance of what Ivan Illich has called the "institutionalization of values" [13]. He argues that health care is one of those areas of life in which modern society teaches us that the things we value are produced only by the "experts": thus, medical and social institutions define our needs and persuade us to purchase their services ([12], p. 406). The result is that we lose our autonomy over health care, which should fundamentally be a matter of self-care.

Furthermore, at the institutional level it is doubtful to what extent we can apply the traditional model of autonomy to treatment decisions made by patients who are institutionalized in critical-care institutions [14]. This doubt is based on the institutional factors that lead patients to be ambivalent about their treatment because they are ambivalent about what others expect of them, or because their circumstances allow them no other way to resolve inner conflicts except by a request to die [5]. Yet it seems incongruous for the decisions regarding reduced autonomy (and consequent treatment decisions) to be made by the very ones who control the institutionalization that reduces or removes the autonomy of those same patients.

Distribution of critical-care medicine. Acting as gate-keepers for hospitals and medical care, the medical profession has long functioned as benevolent redistributors of health-care resources through price discrimination and adjustment of insurance and hospital fees. But as critical-care and high-technology, non-critical-care medicine increase expenses, it is dubious whether the informal "robinhooding" of the parable is a just and rational way of proceeding. Indeed, the very function of beneficence in critical-care medicine becomes problematic when we consider the range of competing views on the ethics of beneficence: first, that beneficence is entirely optional, in the sense that one does no wrong in omitting beneficent acts; second, that there are serious moral duties of beneficence [27]; and third, that beneficence (e.g., Good Samaritanism) can be much more seriously incumbent on political communities than on individual persons ([34], p. 16). An interesting question is whether a commitment to beneficence on the part of the various agents of critical-care medicine (the state, the medical profession, health-care institutions) might not lead us to justifiable moral expectations of health-care interventions that would prevent the very illnesses (including the parable's end-stage renal disease) that bring both citizens and aliens to our critical-care facilities.

Our society is clearly torn between expansive principles of egalitarian justice and restricting factors such as autonomy-based rights and the sheer necessity of cutting costs. This conflict becomes most provocative when it pits citizen against stranger and life-saving actions against other health care.

A strong, legal principle of egalitarianism in emergency medicine was articulated in *Wilmington General Hospital v. Manlove* [35], in which the court held that a private hospital is liable for refusing service to a patient in case of an unmistakable emergency, if the patient has relied on the hospital's custom of rendering aid in such a case - a custom that is presumed from the fact that the hospital operates an emergency room ([7], p. 16). In another important decision, *Guerrero v. Copper Queen Hospital* [9],[5] a court found that a privately-owned hospital is obligated to provide emergency care to all persons who present themselves at the facility for treatment, even in this case, in which the patients were aliens who entered the country specifically to receive the emergency treatment.[6] A similar principle, articulated on philosophical grounds, is that both documented and undocumented aliens have human rights which result from one's humanity and do not depend on citizenship [21].

It is clear that reasonable limits must be placed on the welfare services, such as education or health services, which a nation might grant to aliens, just as the obligation of hospitals to treat all emergency patients regardless of cost must also have a reasonable limit. In resolving these conflicts, it will be necessary to tend to issues of local loyalties and bigotries [8]; theories of national autonomy [2] that could be used to justify a nation's distributing its resources solely on the basis of its autonomy and aside from any considerations of distributive justice; and principles of international distributive justice ([2], [28]).

A major pressure against equality of access to high-cost critical care is the move toward health cost containment [6]. Yet physicians and hospitals have found themselves faced with a proposed policy that would make it a federal offense to withhold food or medical care from infants on the basis of their handicaps [33]. Prominently displayed legal notices in hospital nurseries would list a hot-line telephone number for reporting offenses against medical rights. Thus, in Orwellian fashion, physicians, nurses, and parents were put on notice that Big Brother is watching. This absolute, egalitarian regulation was paired with the same federal government's cut of funds to support the life and treatment of infants who can be saved. Consequently, the absurdities of profoundly conflicting policies threatened to become deeply imbedded in those legal and medical institutions upon which we commonly rely for life-support.

CONCLUSION

A dominant characteristic of critical-care medicine today is the emergence of powerful institutions functioning within a framework of a noncoherent set of values and philosophical perspectives. Thus, anyone who would assign a significant role to the philosophy of medicine for today's era must, it seems to me, take seriously into account not simply the *quandaries* of critical-care medicine, but must also attend to the antecedent values, conflicts, and absurdities that form the ethical issues, as well as the models of ethical response (market ethos, professional ethos, etc.) that indicate which moral principles might be relevant. This broad philosophical task is an urgent one, for critical-care medicine is rapidly molding the moral dimensions of all of medicine. These considerations form the new agenda for the philosophy of critical-care medicine.

Georgetown University School of Medicine
Washington, D.C.

NOTES

[1] This paper was prepared while the author was a Fellow (1982-83) of the National Humanities Center, Research Triangle Park, North Carolina. The author expresses his gratitude to the National Humanities Center, Georgetown University, and the Hillsdale Fund for their support of the project, and especially to those colleagues whose comments were acknowledged in the notes to the antecedent chapter of this volume, 'A Movable Medical Crisis'.

[2] For a perceptive analysis of the privileged status of the medical profession and dissatisfactions with that status, see the article by A. Kenneth Pye [26].

[3] I am grateful to Professor Donald M. Frame for calling this to my attention and for making me aware of Montaigne's ideas of the absurd.

[4] For an incisive analysis of the need for attention to the character of a society as a foundation for social ethics, see Stanley Hauerwas's *A Community of Character* [11], a book written explicitly for the Christian tradition but with philosophical assumptions suggestive of a theory that would be applicable more broadly to social ethics.
[5] This decision of the Court of Appeals of Arizona, Division 2, was vacated by the Supreme Court of Arizona on July 18, 1975, on other grounds [10].
[6] Similarly, the U.S. Supreme Court's 1983 *Plyler v. Doe* decision protects the rights of illegal alien children to a free public education, because of equal protection that must be given to all U.S. *residents* [25].

BIBLIOGRAPHY

[1] Arras, J. D.: 1980, 'Medicine Men, Businessmen', Review of E. R. Brown, *Rockefeller Medicine Men: Medicine and Capitalism in America* (Berkeley: University of California Press, 1979), in *Hastings Center Report* 10: 3 (June), 41-43.
[2] Beitz, C. R.: 1979, *Political Theory and International Relations,* Princeton University Press, Princeton, New Jersey.
[3] Berlant, J. L.: 1975, *Profession and Monopoly: A Study of Medicine in the United States and Great Britain,* University of California Press, Berkeley.
[4] Bondeson, W. B.: 1982, 'Consulting With Integrity: Some Reflections on Team Health Care and Professional Responsibility', in G. J. Agich (ed.), *Responsibility in Health Care,* D. Reidel Publ. Co. Dordrecht, Holland, pp. 185-192.
[5] Burt, R. A.: 1979, *Taking Care of Strangers: The Rule of Law in Doctor-Patient Relations,* Free Press, Macmillan Publishing Co., New York.
[6] Carney, K.: 1981, 'Cost Containment and Justice', in E. E. Shelp (ed.), *Justice and Health Care,* D. Reidel Publ. Co., Dordrecht, Holland, pp. 161-178.
[7] Flannery, F. T.: 1975, 'Hospital Liability for Emergency Room Services – The Problems of Admission and Consent', *Journal of Legal Medicine* **18**, 15-19.
[8] Gorovitz, S.: 1977, 'Bigotry, Loyalty, and Malnutrition', in P. G. Brown and H. Shue (eds.), *Food Policy: The Responsibility of the United States in the Life and Death Choices,* Free Press, Macmillan Publishing Co., New York, pp. 129-142.
[9] *Guerrero v. Copper Queen Hospital,* 22 Ariz. App. 611, 529 P. 2d 1205 (1974).
[10] *Guerrero v. Copper Queen Hospital,* 112 Ariz. 104, 537 P. 2d. 1329 (1975).
[11] Hauerwas, S.: 1981, *A Community of Character,* University of Notre Dame Press, Notre Dame.
[12] Hunt, R. and Arras, J. (eds.): 1977, *Ethical Issues in Modern Medicine,* Mayfield Publishing Co., Palo Alto, California.
[13] Illich, I.: 1975, *Medical Nemesis: The Expropriation of Health,* Calder and Boyars, London.
[14] Jackson, D. L. and Youngner, S.: 1979, 'Patient Autonomy and "Death with Dignity" ', *New England Journal of Medicine* **301**, 404-08.
[15] Kass, L. R.: 1982, 'Professing Ethically: On the Place of Ethics in Defining Medicine', A Conference on the Humanities and the Profession of Medicine, *The Humanities and the Profession of Medicine,* National Humanities Center, Research Triangle Park, North Carolina, pp. 85-100.
[16] Leaf, A.: 'The MGH Trustees Say No to Heart Transplants', *New England Journal of Medicine* **302**, 1087-1088.

[17] MacIntyre, A., 1981: *After Virtue: A Study in Moral Theory,* University of Notre Dame Press, Notre Dame.
[18] Mitchell, C.: 1982, 'Integrity in Interprofessional Relationships', in G. J. Agich (ed.), *Responsibility in Health Care,* D. Reidel Publ. Co., Dordrecht, Holland, pp. 163-184.
[19] Montaigne, M. de: 1588, 1965, 'Apology for Raymond Sebond', in D. M. Frame (trans.), *The Complete Essays of Montaigne,* Stanford University Press, Stanford, California, pp. 318-457.
[20] Newton, L. H. and Pellegrino, E. D. (eds.): 1982, *Collective Responsibility in Medicine,* thematic issue of *Journal of Medicine and Philosophy,* Vol. 7, No. 1.
[21] Nickel, J. W.: 1983, 'Human Rights and the Rights of Aliens', in P. G. Brown and H. Shue (eds.), *The Border That Joins: Mexican Migrants and U.S. Responsibility,* Rowman and Littlefield, Totowa, New Jersey, pp. 31–45.
[22] Norris, J. A.: 1980, 'Current Status and Utility of Emergency Medical Care Liability Law', *Forum* **15**, 377-405.
[23] Nozick, R.: 1974, *Anarchy, State and Utopia,* Basic Books, New York.
[24] Orwell, G.: 1961 (1949), *1984,* with an Afterword by Erich Fromm, New American Library, Signet Classics, New York.
[25] *Plyler v. Doe,* 50 *United States Law Week* 4650 (1982).
[26] Pye, A. K.: 1982, 'A Layman Looks at the White Coats', *North Carolina Medical Journal* **43**, 655-660.
[27] Reeder, J. P.: 1982, 'Beneficence, Supererogation, and Role Duty', in E. E. Shelp (ed.), *Beneficence and Health Care,* D. Reidel Publ. Co., Dordrecht, Holland, pp. 83-108.
[28] Reich, W. T.: 1978, 'Codes, Oaths, and Prayers of Medical Ethics', in W. T. Reich (ed.), *Encyclopedia of Bioethics,* 4 vols., Free Press, Macmillan Publishing Co., New York, Vol. 4, pp. 1721-1815, at 1731-1763.
[29] Relman, A. S.: 1980, 'The New Medical-Industrial Complex', *New England Journal of Medicine* **303**, 963-970.
[30] Shue, H.: 1982, 'The Geography of Justice: Beitz's Critique of Skepticism and Statism', *Ethics* **92,** 710–719.
[31] Simms, V. J.: 1974, 'A Reconsideration of Orwell's *1984*: The Moral Implications of Despair', *Ethics* **84,** 292–306.
[32] Starr, P.: 1982, *The Social Transformation of American Medicine,* Basic Books, New York.
[33] U.S., Department of Health and Human Services: 1983, 'Nondiscrimination on the Basis of Handicap', *Federal Register,* Vol. 48, No. 45 (March 7), 9630-9632.
[34] Walzer, M.: 1981, 'The Distribution of Membership', in P. G. Brown and H. Shue (eds.), *Boundaries: National Autonomy and Its Limits,* Rowman and Littlefield, Totowa, New Jersey, pp. 1–35.
[35] *Wilmington General Hospital v. Manlove,* 54 Del. 10, 174 A.2d 135 (1961).

H. TRISTRAM ENGELHARDT, JR.

MORAL TENSIONS IN CRITICAL CARE MEDICINE: "ABSURDITIES" AS INDICATIONS OF FINITUDE

I. INTRODUCTION

Warren Reich's parable [15] presents us with moral dilemmas because we are free beings with finite capacities and resources, but infinite expectations and hopes. Though we are all doomed to die, and many of us to suffer, we often dream of immortality, including physical immortality. We are reluctant to accept the notion that, as finite beings, not gods and goddesses, we have finite resources. We are, however, pressed to make the painful choices of humans, which gods and goddesses may avoid. We must choose, both as individuals and as societies, which goals and endeavors are worthy of what portion of our efforts and resources: critical care, the arts, the humanities, or the fleeting, fleshy pleasures of this life. If one places most of one's resources in the development of critical care facilities for all who might come, under all circumstances, there will be little left for a good police force, good roads, good restaurants, good bourbon, and good art. Humans must choose among competing possibilities for the investment of their resources.

Since we are finite in our capacities, not only are our resources limited, but our knowledge is limited as well. In responding to risks, individuals can pool resources in order to have them available to cover the cost of a particular genre of expenses likely to be incurred by uncontrollable events such as disease, earthquake, fire, and storm. Through various actuarial calculations one will be able to judge in general what the long-run likelihood of such disasters will be. However, particular individuals will sometimes discover after the fact that they have under-insured themselves against such catastrophes. Planning for the future thus requires us to face the unpleasant fact that we cannot protect ourselves against all possible future losses. At some point, we will properly choose to keep some money for good art and good bourbon rather than investing all in health care and research. As a consequence, we individually, or as groups or societies, will come to be in need of critical care medicine without having paid for sufficient insurance. To put it another way,

J. C. Moskop and L. Kopelman (eds.), Ethics and Critical Care Medicine, 23–33.

when we as individuals, groups or organized societies set aside some, but not all, of our resources for critical care medicine, we are fashioning a particular insurance plan and will, as finite beings, have to recognize its limitations.

This circumstance will force us to the painful but inevitable position of having to remind individuals that it is unfortunate, but not unfair, that some will die because resources are not available to them. For instance, I will argue there is nothing in itself immoral about the circumstance that in the United Kingdom individuals after a certain age do not usually receive renal dialysis through the National Health Service [9]. Such a circumstance is surely unfortunate for those individuals, but it is not unfair in the sense of providing the basis for blaming others for not being forthcoming with additional resources. Their society, a democracy, has fashioned a particular societal insurance program which leaves some funds for ale at the pub, and as a result leaves certain possible losses in the natural lottery uncovered. Insofar as it represents a societal choice, British citizens cannot claim to have been treated unfairly in not receiving health care not covered under the societal "policy."

What Reich terms "absurdities" are to be viewed as circumstances in which our desires outstrip our capacities. Though we might hope to treat all equally with optimal health care, we simply do not have, and in all likelihood will not have, the resources given the range of interests we wish to pursue. Indeed, not all individuals are willing to join in insuring for maximal health care. Certain hopes turn out, then, to be absurd in the sense of involving expectations that ignore our condition as free, finite beings. It is only if one thinks that one can guarantee to all, individual health care that is both *equal* and *the best possible,* and have resources left over for other human goals, that one has embraced a moral and material absurdity.

II. BASIC NOTIONS

Before examining Reich's parable, one must recognize the unavoidable roots of inequalities in life generally, and in health care in particular, including critical care medicine.[1]

(1) First, there is the natural lottery. Some are born with genetic diseases, others with an excellent constitution; some contract diseases through no fault of their own, others do not; some are born in El Salvador, others in the United States of America. The outcomes of the

natural lottery are as such not unfair, but simply unfortunate, unless one is to blame God. They are tantamount to acts of God, they are the deliverances of blind causal chains beyond the immediate influence of human volition.

(2) There is in addition the social lottery. Some individuals receive inheritances, other do not. Some are able to make advantageous agreements with others, and others fail to make such agreements. The social lottery expresses the outcome of human choice including injuries by individuals and groups other than society as a whole. Insofar as persons are free, they will fashion associations, make friends, empower their children and rectify out of beneficence some but not all of the injuries to third parties.

As the result of the social and natural lotteries, some come to health care with great physical problems, others with few; some come rich and others come poor. The natural and social lotteries have as their consequence a third, important derivative source of inequality.

(3) People differ in their entitlements. Such differences are eradicable as long as individuals are free. There may indeed be certain general common entitlements. All may share in a certain claim to the fruits of the earth [1, 11, 13]. However, it is also the case that individuals claim to own things, themselves, their talents, their abilities, and their energies.

A sense of private ownership, and its borders with common ownership, can be appreciated if one looks at the difficulties in generating authority for the allocation of resources in a society. There are four major sources of authority [2, 4]. The first is force. However, an appeal to force will not give a satisfactory intellectual answer to a moral question. The second is the grace of a common moral viewpoint. However, it is unlikely that all will convert to the same understanding of the good life. The third is an appeal to rational arguments. In order to choose the proper view of the good life, one will need to appeal to some notion of a proper moral sense or of proper moral deciders. However, in order to choose the right moral sense or to impute the right moral sense to hypothetical observers or contractors, one will need to appeal to a yet higher moral sense in order to know which moral sense to choose, and so on, *ad indefinitum* [3]. Thus, it becomes quite difficult to establish a particular moral sense as having authority. Yet the issue of moral authority is crucial to public policy planning, insofar as such planning includes the use of coercive force.

Public policy in the 20th century exists in a serious moral vacuum. First, there is no generally shared religious consensus concerning the proper goals of public policy. The loss of that consensus, insofar as it ever existed, was signaled by the Reformation, Renaissance, and finally the Enlightenment, which produced numerous competing views of the proper ways of shaping public policy. Further, the Enlightenment was not able to produce a substitute for religious consensus by framing an authoritative moral viewpoint based on the arguments of reason alone. Yet such arguments are required for those who would, through coercion, reallocate privately owned goods and private energies. The burden of the proof is upon the individual who claims to know what is right. Those who make no claims assume no such burdens. That burden becomes doubly severe when an individual claims singly or in a group (even if that group terms itself a state) to have the authority to impose a particular viewpoint by force.

A resolution of this problem has in part been found in certain libertarian democratic approaches. If authority cannot be derived from the grace of God, or from a successful rational argument, it can be derived from a fourth source, the consent of all involved. When authority cannot mean the authority of God or the authority of the only rational way of acting, it can at least mean the authority of those joined in a project. However, absent a special rational, perhaps metaphysical argument of some sort, the last genre of authority must be actually conveyed by actual persons. One does not have the authority of a group unless the members of that group have, in fact, agreed to convey authority. Still, whenever rational arguments cannot in principle establish a concrete view of the good life or of the proper allocation of resources, one can at least create through common consent an acceptable approach. Even if one cannot discover a proper pattern for public action, one can unpack the grammar of the notion of a peaceable community as an alternative to the resolution of disputes through force. Insofar as one is interested in a way of resolving moral disputes without force and with authority, there still remains a mode secure against the collapse of the Judeo-Christian synthesis and the Enlightenment dream of being able to discover an authoritative moral sense. One can still make moral claims with authority and mean something by these claims that can be generally justified: acting in a contrary fashion would make impossible the peaceable resolution of disputes with authority. The minimum notion of morality, and therefore the most generally justifiable notion of morality, is that of the peaceable community where unconsented-to force is not used against others who similarly eschew such force and where therefore moral disputes can be resolved by peaceable negotiations, including rational arguments.

This fourth and final root of moral authority, though, places the individual as a central source of societal authority. It provides, as a consequence, a principle in terms of which private entitlements can be recognized, however limited. Basic rights under a libertarian approach will concern those areas of conduct where others cannot show authority for their use of coercive force. Thus, viewing pornography in the privacy of one's home is a fundamental or basic right in the sense that absent (1) the general grace of God or (2) a conclusive moral argument to establish the immorality of such actions, or (3) an agreement of the persons involved not to engage in such activities, it will not be possible to show that others have authority to stop persons from engaging in such actions. Past agreements of one's ancestors to particular laws or to a constitution will not morally restrict such basic rights unless there is a conclusive rational argument for a special genre of hereditary slavery to the effect that if one's ancestors bartered away their rights under a constitution, they were able to barter away at the same time the rights of their descendants. Such an argument is highly metaphysical and in the end unsustainable.

These considerations have important implications for understanding the standing of private and common properties. On the one hand, it will be very difficult to show that anyone has the authority to constrain the energies and talents of others without their consent. Such constraint would violate the minimal condition of a peaceable community. There will, as a result, be talents and energies, and products of these talents and energies, that will be primordially private in this sense. Products of conjoint labors, including the endeavors of society, may also in this fashion be seen to belong to particular societies. However, it will be very difficult to establish ownership of land and things insofar as these are not in one way or the other rendered products of human talents and energies. In any event, such transformations will not render things or land completely the property of either individuals or groups. They will rather have property rights in such things or land. Everyone, however, will retain an equal right in the unconverted residuum, that element of things and land which has not been rendered a product of the energies of individuals or groups.

It is considerations such as these that have led individuals to argue for a sort of general negative income tax, based on the rent due to all for the use of material that cannot reasonably be understood to have been rendered wholly private. One thus finds special adaptations of Locke by

Ogilvie [11], Paine [13], and Brody [1]. These reflections have many implications. One of the most important for us here is that some entitlements remain irradicably private.

Once one recognizes some private entitlements, one is faced with inequalities, for free people will choose according to diverse criteria as to how they will dispose of what is theirs. In addition, insofar as rational arguments fail to disclose *the* view of the good life, one will be forced to tolerate numerous views of the good life. The right to associate with others in pursuing one's view of the good life is simply a reminder that others do not possess the authority to use force to stop such associations [5]. Persons will, then, have the right to fashion special associations with special insurance policies and special benefits.

Thus, given the natural and social lotteries, the existence of at least certain private entitlements, the freedom of persons and the absence of *an* authoritative view of the good life [2], persons will have different and unequal entitlements. Many of the inequalities will not be unfair in the sense of someone being deprived of something that is his or her due, but simply unfortunate. In other words, there will be inequalities that will not count as inequities. There will as a result be more than one tier of health care. Though there may be a common tier of entitlements, there will be other tiers available only to those with funds. Another way of putting this is the following: though a society may from its common resources purchase a certain level of insurance with respect to health care needs, it cannot cover all of the losses of the natural and social lotteries. Further, particular individuals and groups may take out special insurance policies of their own, or simply set aside enough resources for some of their losses. Moreover, individuals can use their private goods to barter for the services of others. One finds this even in Soviet bloc countries, where individuals more or less openly purchase better service on the side than the established system is able to deliver. In short, the black market is morally instructive and individuals have in circumstances such as this an inalienable right to participate in the black market.

The circumstances of critical care medicine will constrain those involved to recognize that the very best of everything cannot be provided for everyone. We will need to draw the line between what will be unfair distributions of health care, and simply unfortunate levels of health care, the inevitable outcomes of finitude and freedom. Unfair distributions will be those which provide less than that amount of health care to which individuals are entitled. Unfortunate distributions will be those which do not meet all of the health care needs of individuals, but which yet provide all the care to which the individuals are entitled.

III. INTERNATIONAL, NATIONAL, AND PRIVATE INSURANCE POLICIES

One might argue that all should be provided with a minimum decent level of insurance against losing at the natural and social lotteries [6]. This one might conceive of through the metaphor of an international health insurance policy to be provided for all. However, it is very unlikely that successful arguments will be available to justify coverage of all for all expensive forms of treatment under such a policy. If Brody [1] is right, one would have only those funds available for common international investment in such a policy that were the fruits of international conjoint endeavors. The rest would be private funds available from what would be tantamount to an international negative income tax, which could be invested only as the recipients wished.[3] In any event, national forms of coverage, as for example through Medicaid and Medicare, cannot cover, nor should they cover, all possible losses through the natural and social lotteries. Such coverage would be impossible. The attempt to provide it would undoubtedly lead to confiscating privately owned goods in order to pursue an ephemeral goal. Finally, in addition to international and national insurance policies, private insurance policies will develop, as currently exist even in the United Kingdom.

These reflections lead to the inevitable moral conclusion that hospitals have no moral obligation, legal issues notwithstanding, to treat all who come to their doors, especially when such treatment will entail considerable costs. This view may in part be contrary to court holdings such as *Guerrero v. Copper Queen Hospital* [7] which have supported the duty of private hospitals to provide emergency room care, indeed in that case, to Mexican citizens. Such holdings, however, reflect an ethos insufficiently sensitive to the limitations of scarce resources. And as the character of private hospitals was changed in Arizona through state regulation, according to *Guerrero v. Copper Queen Hospital,* further changes are likely as well. The influential 1961 case of *Wilmington General Hospital v. Manlove* argued that the duty to treat was based on individuals "relying on [an] established custom" [7]. And as such cases indicate, customs change. In fact, the moral bases for such changes can be envisaged with regard to very expensive critical care medicine. Responses to these limitations will likely include setting limits on the obligations of private hospitals to provide extremely costly care. They may also include setting limits for public or private hospitals when El

Salvadorans come to their doors as described in Reich's parable. One will need to determine with care the terms of the various insurance policies one possesses.

Finally, given the role of freedom in the moral life, there will be nothing wrong with individuals forming an association to deliver health care for profit. No force is used against the innocent, and a social good is pursued: quality health care. Indeed, such companies may be of benefit not only to the greatest number, but indirectly to the least well-off classes through the development of innovative means of health care delivery and treatment. However, such companies will face conflicting goals. Such endeavors will need on the one hand to provide good treatment and serve their communities, and on the other hand to make a profit. Since making a profit is the necessary condition for the continued existence of a company for profit, considerations of profit will necessarily, and properly, be central. Still, if no one is forced to be admitted to such hospitals, there is no fundamental moral reason against their setting requirements for payment, or regarding levels for service. Hospitals for profit are an expression of the freedom of humans to fashion associations. Indeed, public attempts to legislate them out of existence by imposing severe financial burdens would be an immoral act of majoritarian force.[2]

Finally, to fashion one kind of association, rather than another, destines one and one's associates to be able to do some things with ease, and others only with difficulty. One cannot be free to do everything with equal ease. *Omnes determinatio est negatio.* To choose one thing is to give up other things. Associations are not intrinsically Orwellian, or coercive. They are expressive both of human finitude and freedom.

IV. THE PARABLE RE-EXAMINED

Given these distinctions, we can develop some responses to Reich's parable. First, we must observe that it teaches us the importance of becoming clear in advance about the ways in which we should treat problematic cases. The absurdities he characterizes are the products of moral unclarity and lack of forethought and planning. As we develop more costly treatment, we will need to develop humane modes of rejecting or transferring patients who come to hospitals without proper insurance policies. Reich's EmergiMedVan appears on these grounds to be an instructively virtuous compromise. It reflects a beneficent attempt to provide adequate treatment, while transferring individuals to institutions covered by their insurance policies. Here I include metaphorically under the rubric of "insurance policy" those forms of treatment provided free to the indigent by eleemosynary hospitals.

Now to the parable itself, Pilt is right. Whether or not Texans understand the Bible better than Carolinians, they at least, within this parable, understand the counsels of finitude better. Dr. Hilaire Weeks may not give away the company's resources in the support of the noble causes she embraces without in the end having to answer to the company's stockholders. If she had acted other than she did in seeking the company's agreement to engage, both by itself and through charitable contributions, in further funding of a Good Samaritan policy, she would have been stealing, albeit for "noble" ends. Though an attendant at a filling station owned by Big Oil may be moved by compassion to pump gas into the cars of the indigent, it is still wrong. Whether one is Bonnie or Clyde, Robin Hood, or Dr. Hilaire Weeks, one may not transfer goods from those who own them to those who do not, without permission of the owners. One must instead engage in the much more onerous endeavor of gaining the free consent of members of a community for the use of their commonly owned resources in acts of beneficence, including the provision of more encompassing general insurance policies.

As to consent in the EmergiMedVan, I will accept for the purposes of my remarks that it is the case, as Reich states, that "it will not be possible to consult carefully with the patients on board nor with their next-of-kin" ([15], p. 8). *If that is the case*, and since the EmergiMedVan appears to be a decent response to individuals who come to a hospital without strict moral entitlements to its care, the computer program to direct care appears to be an acceptable strategy. It is an attempt to respond as best one can under circumstances where one's beneficence has been coerced. It is surely not an optimal response. Still those who constrain others to provide care to which they have no strict claim cannot protest if they do not receive an optimal response to their wishes. It is enough that they are transferred safely to an institution which has contracted to provide the services to which they are entitled. Since Reich raises the issue of treatment in the EmergiMedVan only in passing, however, I can here but respond in passing, in an attempt to integrate this element of the parable within the general issue of responding to individuals in need of costly care who do not have strict entitlements to that care.

V. SUMMARY

In part I have agreed and in part I have disagreed with Reich. I have agreed absurdities can be engendered by failing to acknowledge human finitude and freedom. It is not that the world is inexplicable. It is rather that we render the world explicable through our conceptual construals. Nowhere does this succeed more than when we fashion the social world. The social world is the product of thoughts and conceptual distinctions. If carefully wrought, it is far from absurd. It will, however, require for its successful fashioning acknowledging the character of human finitude and freedom. Moreover, in order to avoid painful confusions due to individuals arriving with unclear entitlements, we must as far as possible reflect on the kinds of assurances we will give regarding the levels of care provided in various circumstances. Part of treating people fairly is making clear to them what care they can expect. That is not Orwell's theme in *1984.* Orwell was concerned with a web of coercive constraints on freedom, not a web of free choices made in the face of finitude [12].

Given the cost of critical care, we will undoubtedly, as in the United Kingdom, need to recognize that there will be limits on the provision of expensive medical life-saving treatment. This has already been the case with regard to heart transplantation [8]. There is no moral evil in this acknowledgement. On the contrary, it is a step towards the virtues required of finite beings. Of the many moral issues raised by critical care medicine, Reich has surely addressed one of the central quandaries. However, its solutions do not abide in absurdity. The solutions lie in understanding that we as human beings cannot set aside all of the unfortunate circumstances which result from the outcomes of the natural and social lotteries. It is our finitude that limits us with regard to the natural lottery. We will never have all of the life of quality that all of us might desire. Moreover, freedom sets limits upon our ability to set aside the deliverances of the social lottery, which lottery in part expresses the free choices of men and women regarding whom they will love and befriend.

Baylor College of Medicine
Houston, Texas

NOTES

[1] In this section, I forward ideas in part drawn from, and which in part contrast with, the work of John Rawls [14] and Robert Nozick [10]. For a discussion of the application of theories of justice to issues of health care, see [3] and [17].

[2] For a discussion of the limits of societal authority see [3].

[3] I here alter the nation and society-specific elements of Brody's argument.

BIBLIOGRAPHY

[1] Brody, B.: 1981, 'Health Care for the Haves and Have-nots: Toward a Just Basis of Distribution', in E. E. Shelp (ed.), *Justice and Health Care,* D. Reidel Publ. Co., Dordrecht, Holland, pp. 151-159.

[2] Engelhardt, Jr., H. T.: 1982, 'Bioethics in Pluralist Societies', *Perspectives in Biology and Medicine* **26**, 64-78.

[3] Engelhardt, Jr., H. T.: 1985, *Bioethics: An Introduction and Critique,* Oxford, New York.

[4] Engelhardt, Jr., H. T.: 1980, 'Personal Health Care or Preventive Care: Distributing Scarce Medical Resources', *Soundings* **63** (3), 234–256.

[5] Engelhardt, Jr., H. T. aand Malloy, M.: 1982, 'Suicide and Assisting Suicide: A Critique of Legal Sanctions', *Southwestern Law Journal* **36,** 1003–1037.

[6] Fried, C.: 1976, 'Equality and Rights in Medical Care', *Hastings Center Report* **6**: 1 (February), 29–34.

[7] *Guerrero v. Copper Queen Hospital,* 112 Ariz. 104, 537 P. 2d 1329 (1975).

[8] Health Care Financing Administration: 1980, 'Exclusion of Heart Transplantation Procedures from Medicare Coverage', *Federal Register* **45** (August 6), 52296-52297.

[9] Newman, B.: 1983, 'Frugal Medical Service Keeps Britons Healthy and Patiently Waiting', *Wall Street Journal* **LXXI** (28), 1, 21.

[10] Nozick, R.: 1974, *Anarchy, State, and Utopia,* Basic Books, New York.

[11] Ogilvie, W.: 1781, *Essay on the Right of Property in Land,* J. Waller, London.

[12] Orwell, G.: 1966, *1984,* Harcourt-Brace, New York.

[13] Paine, T.: 1798, *Agrarian Justice,* T. G. Ballad, London.

[14] Rawls, J.: 1971, *A Theory of Justice,* Harvard University Press, Cambridge.

[15] Reich, W.: 1985, 'A Movable Medical Crisis', in this volume, pp. 1–10.

[16] Reich, W.: 1985, 'Moral Absurdities in Critical Care Medicine: Commentary on a Parable', in this volume, pp. 11–22.

[17] Shelp, E. E. (ed.): 1981, *Justice and Health Care,* D. Reidel Publ. Co., Dordrecht, Holland.

[18] *Wilmington General Hospital v. Manlove,* 54 Del. 15, 174 A. 2d 135 (1961).

WARREN THOMAS REICH

"CONCEPTUAL CONSTRUALS" VS. MORAL EXPERIENCE: A REJOINDER[1]

Professor Engelhardt's response [1] is constructive, and the reader may find the differences between our approaches instructive.

I. METHOD

For Engelhardt, the parable of EmergiMedVan simply presents one dilemma (the allocation of critical-care medicine), which we then must resolve by rational analysis, whereas I, in the parable, wanted to look beneath the surface of our critical-care system to dimensions of attitude and culture where the task was not so much to call attention once again to a standard problem of bioethics as to find a massive interplay of moral puzzlements and give voice to them. For there are situations in which only a narrative can adequately probe the *nature* of a moral problem that might too easily be taken for granted in the customary, "case"-oriented, analytic approach.

Engelhardt suggests that the parable's quandaries cede to conceptual construals and ethical principles, whereas I wanted to suggest in the parable and in the main "absurdity" theme of the commentary, that there is a complex setting of moral behavior in critical-care medicine that requires more than the application of "conceptual construals" and ethical principles. For the narrative portrays moral experiences - commitments, conflicts, attitudes and behavior - which, more than conceptual construals, are the stuff of ethics [3]. Thus, as I think the parable suggests, critical-care medicine (like much else in morality) presents us with moral issues that cannot simply be settled deductively by a priori "conceptual construals."

The parable portrays not only a variety of moral experiences but such a conflict among them that they would seem to require a fresh examination from a moral perspective. In this sense the parable-with-commentary is a keynote, a setting of the stage, rather than a definitive staking out of a position. Thus, one might ask: Does the desperate search for life by a few aliens who unnerve us by taking advantage of the one egalitarian door to our medical system in any way suggest that our

J. C. Moskop and L. Kopelman (eds.), Ethics and Critical Care Medicine, 35–40.

current assumptions about membership – membership in our country, membership in insurance systems – is the only valid moral claim to critical-care medicine? Do the integrity and moral commitments of a profession count for anything as huge, new institutions of health care pursue their own agenda that may reshape or repress those moral realities? Isn't there something benevolently incongruous about the pervasive robinhooding among patients by hospitals and hospital corporations to pay bills and to realize a profit? How do the benefits of efficiency, productivity, and profit in health-care institutions compare with the moral character of those very institutions and of the individuals whose character they shape?

It is interesting to note that in Engelhardt's carefully developed theory of health-care allocation, which relies on a libertarian construal of property allocation, moral character and professional responsibility play no significant role. It is also instructive to note that the parable's satiric fantasy of an EmergiMedVan oribiting among hospitals – in a circuit of indefinite duration – meets the requirements of Engelhardt's conceptual construal of autonomous, contracting (but contractually impoverished) individuals.

This points, once again, to our sharp differences, for Engelhardt's quite literal reading and philosophical analysis of the parable might cloud the fact that a parable is itself a distinctive form of ethical discourse. By portraying one or more "contrast experiences," it leads us to ask whether we want things that way, and by proposing a certain sort of moral behavior in response to that situation, the parable suggests what sort of character we or our society might actually become - or might *not* want to become. In addition, a parable can shape our perception of our own moral agency vis-a-vis this particular segment of the moral life. Thus, assuming some moral hermeneutic for understanding the role of narrative in ethics [3], a parable and its components might have a role in the ethics of critical-care medicine very different from Engelhardt's perception of the same.

II. MORAL ABSURDITY

The question of moral absurdity, as well as the Orwellian overtones to moral absurdity in critical-care medicine, may not be the most crucial aspects of my presentation, but they are integral to an understanding of my claims about the nature of the moral problem we are facing in critical-care medicine.

Engelhardt proposes that what I term absurdities are not absurdities at all, but circumstances in which our desire to treat all equally with optimal health care outstrips our finite resources and conflicts with other choices. Thus, my "absurdities" are nothing but "the products of moral unclarity and lack of forethought and planning." In the terms of Engelhardt's explanation, absurdities are less likely in the social world that we produce by our careful, conceptual construals.

But the absurdity I was attempting to portray and then explain was not the same as *dilemmas* about distribution of expensive and scarce critical-care resources (though certainly our society's mismatched views on distribution of health-care resources portray an *incongruous* - and, to that extent, absurd - moral picture). Rather, the moral absurdity is found, above all, in the incongruity of the situation to which EmergiMedVan is a response. For it is profoundly incongruous that a people that takes pride in its rationality, autonomy, freedom, and self-control should *simply discover*[2] that an area of life fraught with intensely felt values and disvalues - health and life, especially in emergency, critical-care situations - is rapidly being dominated by new institutions that have acquired power over health-care distribution in a way that radically alters the ethos of medicine, the roots of professional ethics, what it means to be a sick person in our society, and above all, what it means to be a stranger in our society.

Furthermore, new corporate entities, while inserting health care into a market mechanism subordinate to profit considerations, radically alter the autonomy of the health-care consumer, the health-care provider, the corporate health-care professions, and many health-care institutions. Those most deeply affected by the shift are not participants in effecting the change, and it is not clear that they share the moral beliefs needed to assure their acceptance of the moral, political, and economic implications of the move. Finally, it is absurd that desperate aliens on the verge of losing their lives through reversible health conditions should have to play tricks on our health-care system before we are made sufficiently aware of the need to reflect and act on what it is that we owe to strangers, whether they be aliens or citizens, as well as the need to set limits to our justice and beneficence.

Now those are the ingredients for a profound sense of incongruity and alienation, at least among a people who cherish autonomy and are committed to a variety of values that are in conflict with those selected for ascendancy by the new powers. They are the ingredients not only for a sense of incongruity but of the ridiculously incongruous, which is the definition of absurdity.

Engelhardt obviously sees no parallel with Orwell's *1984* in the parable, partly because he perceives the parable as the presentation of a mere dilemma that requires conceptual clarity and a rational solution. That solution entails acceptance of the fairness of not getting medical care (at least beyond some decent minimum) if you haven't freely chosen to pay for it, e.g., in an insurance-type association. "Orwell was concerned with a web of coercive constraints on freedom," Engelhardt points out, "not a web of free choices made in the face of finitude."

Whatever I may say about Orwell will have no meaning if the parable is submitted to Engelhardt's rationalist reduction. But if one were to return for a moment to the disjointed experience of the absurdity that I described above and portrayed, I hope, in the parable, one might see some parallels with Orwell's moral tale. At least, the attempt to make the parallel might be instructive.

The parallel I was drawing was not with Orwell's theme of dehumanization by a political elite through terror and repression, but with his lesson about gradual dehumanization in a bureaucratized society. According to Erich Fromm, Orwell implied that the danger of dehumanization and alienation exists not only in the totalitarianism of Hitler and Stalin, which were such vivid threats when he wrote in the 1940's, but "that is a danger inherent in the modern mode of production and organization, and relatively independent of the various ideologies. Orwell, like the authors of the other negative utopias, is not a prophet of disaster. He wants to warn and to awaken us . . . ." ([2], pp. 267, 257). Thus, the parallel between the world of EmergiMedVan and Orwell's *1984* is suggested by Fromm's words: "George Orwell's *1984* is the expression of a mood, and it is a warning" ([2], p. 257). We have long been aware of the dehumanization of the person through bureaucratization of individual institutions, including health-care institutions, but have not thought about the sudden loss of power of individual institutions to conglomerate corporations. It is no minor consideration that the moral and socioeconomic impact of such shifts in power are cushioned by the doublethink of "assuring the quality of health-care," through actions taken "in the best interests of the patient," in an effort "to reduce costs." We simply do not know what meaning those maxims bear in this context.

Thus, EmergiMedVan is a symbol of the absurdities that will be utilized by those intent on retaining power and profit behind the precarious mask of professional and economic correctness. Those are the Orwellian overtones I would like to spotlight in our present health-care trends.

III. POINTS OF CONVERGENCE

There are several points on which Engelhardt and I agree. First, being conceptually clear about the limitations of the world in which we live may well diminish the likelihood of experiencing moral absurdities (yet, in my view, no conceptual construals have the power so to arrange our moral perceptions as to prevent moral absurdities, quandaries, and dislocations of all sorts). Second, we agree that the solutions to the central quandaries of critical-care medicine do not *lie* in absurdity. (This is true, in my view, for we must and can be about the business of assessing and controlling power, developing rational responses to moral problems, and sketching out the requirements of character for individuals and societies.) This view that we share undercuts an important element in Orwell's *1984*, for while there may be an aspect of powerlessness in the scenario that I see, there is not one of hopelessness. Third, rational analysis and argument in response to the dilemmas of critical-care medicine are essential elements in (medical) ethics (yet not so fundamental, in my view, as the turn to moral experience and the models that give shape to our principle-based discussions).

Georgetown University School of Medicine
Washington, D.C.

NOTES

[1] This paper was prepared while the author was Visiting Research Professor for Bioethics at the University of Tuebingen, Federal Republic of Germany. The author expresses his gratitude to the Alexander von Humboldt Stiftung for its senior professor award during the 1983 spring semester.

[2] What was so urgent about Relman's analysis of the "new medical-industrial complex" was precisely the "unheralded" and unexamined rise of this huge industry ([4], p. 963).

BIBLIOGRAPHY

[1] Engelhardt, Jr., H. T.: 1985, 'Moral Tensions in Critical Care: "Absurdities" as Indicators of Finitude', in this volume, pp. 23–33.
[2] Fromm, E.: 1961, 'Afterword', in *1984*, by George Orwell, New American Library, Signet Classics, New York.
[3] Mieth, D.: 1977, *Moral und Erfahrung,* 3rd edition, Verlag Herder, Freiburg and Vienna.
[4] Relman, A. S.: 1980, 'The New Medical-Industrial Complex', *New England Journal of Medicine* **303,** 963–970.

JAY KATZ

CAN PRINCIPLES SURVIVE IN SITUATIONS OF CRITICAL CARE?

Law's doctrine of informed consent, now twenty-six years old, has as yet not significantly improved the quality of physician-patient decision making. The doctrine, however, has engendered an unsettling and unsettled debate about the respective rights of physicians and patients to make choices for and with one another. That informed consent's fate would for a long time remain uncertain was foreshadowed in *Natanson v. Kline*, the first opinion to construe in some depth this novel common law doctrine. Justice Schroeder, after boldly asserting that patients' rights to self-decision were to be found in Anglo-American law's "premise of thorough-going self-determination that [considered each man] to be master of his own body" [12], departed all too quickly from this premise by allowing physicians to retain considerable, if not sweeping, authority over patients' medical lives.

Informed consent's uncertain fate was also foreordained by the historical fact, frequently overlooked, that inviting patients' participation in decision making is an alien prescription for physicians to dispense. Obligations to invite patients to share the burdens of decision have roots neither in the Hippocratic Corpus nor in contemporary Principles of Medical Ethics, beyond the most recent terse acknowledgment that informed consent in this day and age constitutes *social* policy [1]. Whether social policy will become medical policy only time can tell.

Thus, it is not surprising that the traditional silence that has pervaded decision making between physicians and patients has not been substantially improved during the last twenty-six years. Law's doctrine has not led to respectful conversation, to mutual sharing of the burdens that medical choices entail. The inhuman consent forms and physicians' monologues on risks, spawned by informed consent, attest to this fact. The persisting unbridged gulf between the idea of informed consent and its application in practice must be kept in mind, otherwise the enormity of the task to translate disclosure and consent into viable principles will not receive the caring attention it deserves.

Yet, the doctrine of informed consent has introduced the principle of

J. C. Moskop and L. Kopelman (eds.), Ethics and Critical Care Medicine, 41–67.

self-determination into medicine, a principle that eventually could revolutionize the feudal medical practices in which physician-patient decision making is still embedded. Self-determination is one of those "self-evident" propositions that, once having been proclaimed, may have to be given its due. Its powerful and persuasive appeal cannot be denied forever.

In this essay I have assigned myself a paradoxical task. I shall seek to advance the cause of self-determination by exploring some of the situations in which patients' refusals to undergo diagnostic tests or to submit to treatment should not necessarily be honored. I intend to demonstrate that forcing ourselves to think critically about the reasons for *rare* exceptions to the rule - respect for patients' final choices - to which I am otherwise committed, will clarify and strengthen physicians' general obligations to inform patients and act only with patients' consent. Indeed, it may turn out that clearly defined exceptions will enhance, rather than undermine, the principle of self-determination.

Since I shall argue for what at first blush appears to be a disrespect of a principle that deserves greater caring attention than it has received in medical practice, I must comment first on some of the problems that adherence to principle poses for interactions between physicians and patients. I shall then turn to a clinical example, a physician's dramatic encounter with a patient during which the doctor honored his patient's unwillingness to undergo diagnostic tests, even though death would be a likely consequence. The analysis of their interactions will illuminate much about my final task, the exploration of the questions: When, and why, should patients' wishes and choices not be honored by physicians?

Finding answers to these questions is crucial not only for situations of refusal but also for the more common situations in which physicians and patients seemingly reach agreement about a proposed intervention. The fear of potential refusal has cast a dark shadow over all physician-patient deliberations. Physicians' fears over the possibility of patients' refusal or unwillingness to agree to the intervention preferred by their doctors, if treatment alternatives, – their risks, benefits and uncertainties – are frankly acknowledged, has made a contribution all its own to the manipulation of disclosure and, in turn, of patients' consent. Thus, what constitutes "refusal" and what constitutes "consent" become difficult to identify unless one wishes to equate spurious consent with meaningful consent.

How to proceed in the face of disagreement has always troubled physicians, and particularly so in the age of informed consent. Greater clarity about when patients' wishes and choices should be either disregarded or respected could contribute to lessening the pervasive engineering of consent [4] and, instead, to facilitating a more open confrontation of differences of opinion.

I. THE TYRANNY OF PRINCIPLE

Stephen Toulmin in a recent article has reminded us once again "that a morality based entirely on general rules and principles is tyrannical and disproportioned, and that only those who make equitable allowances for subtle individual differences have a proper feeling for the deeper demands of ethics" [16]. He has identified here the eternal tensions between rules and principles (indispensable for the control of ill-considered personal and professional judgments), on the one hand, and discretion (dictated by the complexities of individual situations and needs), on the other. These tensions cannot be eliminated. They deserve attention so that neither principle nor individual judgment do violence to decision-making processes.

I agree with Toulmin that if one had to choose between "casuists," committed to case-by-case analysis, and "moral enthusiasts [bent on nailing] their principles to the mast" ([16], p. 38), the former are to be preferred over the latter. The case for principles and rules, however, can be strengthened once we more fully consider other problems posed by "the abstract generalizations of theoretical ethics" ([16], p. 31). I shall comment on two of these problems.

The first problem resides in the chasm that separates abstract principles from the specific, concrete, and highly individualized situations in which they need be applied. John Ladd, among others, has pointed to this problem and quoted John Stuart Mill in support of the need for "*secondary principles* to mediate between the abstract super-principle and concrete cases of action or decision-making" ([10], p. 2). While a problem, it is not an insurmountable one. To better manage it, bioethics must develop – as is equally true for my own discipline, psychoanalysis – a clinical theory that mediates between abstract concepts and real-life cases.

Contemporary bioethics suffers from the lack of "a clinical theory." Instead, Kant, Mill, and others have been invoked too readily to give answers to clinical problems that their abstract formulations are ill-suited to provide. These philosophers at best can provide a "Weltanschauung," yet one that in the course of translation to practical cases generally becomes wittingly and unwittingly infiltrated by idiosyncratic personal and professional preferences that may say more about the commentators and their orientation than the principled problems before them. If ethicists avoid that trap by relying solely on abstract principles to guide their recommendations, their ultimate answers tend to defy human realities. "The tyranny of principles," a felicitous phrase employed by Toulmin, highlights the inhuman demands principles can make as guides for human conduct. As inquiries into ethical decision making mature, clinical theories need to be developed to reduce this dilemma.

The second problem resides in the failure to distinguish between the moral guidance principles provide for regulating the interactions between individuals and the views they imply about human capacities to so conduct themselves. These issues require separate consideration. Take autonomy as an example. It is spoken of both as a moral principle, a *right* possessed by persons that safeguards them from unwarranted interference by others, and as a *capacity* that persons possess for the exercise of such a right. While the two interrelate, they must be considered separately, if only because doing so will compel a clearer recognition of the extent to which one's views on capacity for autonomy either influence or should influence one's view on a right to autonomy.

Thus, any disagreements on the respect that should be accorded to patients' "inalienable right" to autonomy ultimately may be grounded in different underlying assumptions different commentators make about *all* human beings' capacities for rational and voluntary thought and action, and not merely those who "are substantially nonautonomous, whose actions are essentially nonvoluntary"([17], p. 195). This problem can be better addressed by making conceptual distinctions between rights and capacities. Such distinctions will underscore the significance of human beings' psychological nature as an independent guiding concept for decision making.

Before exploring this problem, I would like to comment briefly on some intriguing observations of Toulmin's that bear on this issue. In his article, "The Tyranny of Principles', he observed that the assertion of "ethical objectivity has led . . . to an insistence on the absoluteness of moral principles that is not balanced by a feeling for the complex problems of discrimination that arise when such principles are applied to real-life cases" ([16], p. 31). He deplored the concomitant "distrust of individual discretion" and found the reason for such distrust in the ascendancy of an "ethics of strangers" over an "ethics of intimacy." In the former, "respect for rules is all, and the opportunities for discretion are few" ([16], p. 35), while in the latter these conditions are reversed. He finally expressed his "frail social hope . . . that the ethics of discretion and intimacy can regain the ground it has lost to the ethics of rules and strangers" ([16], p. 37).

While I do not question the need for discretion, I am not as sure as Toulmin seems to be that an ethics of intimacy is to be preferred. The supposed intimacy of physician-patient relations has led to considerable abuse of their fiduciary relationship and there is much to be said in favor of a greater appreciation by both parties that they are, and will always remain strangers to one another or, at a minimum, both strangers and intimates.

I find an explanation for Toulmin's observation that systems of abstract principles define ethical relations between strangers in another property of abstract principles. Since they are frequently formulated without regard to the psychological nature of human beings, abstract principles sound foreign to the heart, if not to the mind. I do not believe that abstract principles require such a disregard of human psychology. Indeed, they never do, for a careful scrutiny of any principle reveals all kinds of hidden, albeit woefully mutilated, assumptions about human nature.

For example, Immanuel Kant, in artificially restricting his conception of autonomy to capacities to reason, without reference to human beings' emotional life and their dependence on the external world, projected a vision of human nature that estranged his principle from human beings and from the world in which they live. That Kant did so deliberately and with full awareness, because his agenda necessitated such a strategy, is a separate matter. What does matter is that Kantian principles, and those of others as well, make demands for human conduct that human beings cannot fulfill. The striking discrepancy between Kant's principles and his clinical examples makes this quite clear ([8], p. 30). The examples are hospitable to the human discretion Toulmin advocates, the principles are not. Thus, abstract principles that do not take human nature more fully into account estrange by creating too great a gulf between principle and man. In the last part of this essay I shall explore the interrelationship between the principle of self-determination and psychological autonomy in order to illustrate that such a fate is not an inevitable consequence of principles.

If I am proven wrong, it must at least be recognized that all abstract principles contain implicit assumptions about human nature, and that these assumptions differ, depending on the commentator. These assumptions need to be made explicit. It may then turn out that it is not the principle that makes application to practical problems so difficult - although this will remain a problem in the light of the ever-present interplay of competing principles - but rather the unreal and estranging psychological assumptions that can inhere in principles. The debate over principles and their relevance to the resolution of human problems will then encompass a new dimension and pose new questions: To what extent do they comport with the biological, psychological and social nature of human beings? To the extent they do not, what is the relevance of such principles as guides to human conduct?

Different theories about the nature of human beings will of course give different contours to principles, and the theory I shall propose, based on psychoanalytic propositions, is not necessarily the best one. I do not wish to plead its acceptance, if only because like all theories, it is not value free. I only wish to suggest that we must commit ourselves to a psychological theory and that whichever theory we embrace must be more self-consciously articulated in order to assess its impact on principle and, in turn, its relevance as a guide to understanding human interactions. Thus, the controversy over the respective value of "abstract generalizations of theoretical ethics" or of "a sound tradition in practical ethics," that as Toulmin observed [16] can lead adherents of the former to "oversimplification" may reside not so much in the inevitable oversimplification inherent in principle but in the flawed nature of the assumptions about human bio-psychology that principles so frequently make. Put another way, the tyranny of principle that Toulmin so correctly decried, may reside not only in the "absolutists' [denial of any] real scope to personal judgment" ([16], p. 31), but also in the inability of human beings to conform to principles based on the apsychological nature of principles constructed by absolutists.

II. MR. D.

A clinical encounter between a remarkable physician, Mark Siegler, and his patient, Mr. D. [14], bears witness to the agony of decision making when doctors and patients disagree. Siegler ultimately decided to honor his patient's wish not to undergo diagnostic tests, even though Mr. D.'s refusal might endanger his survival. For a physician to so conduct himself is itself a rare event. What Siegler then did, however, is even rarer in the annals of medicine. He tried to explicate to himself and then to the world at large the reasons for his decision. His courage not to hide behind a veil of silence, once having acted on his beliefs, provides me and others with the opportunity to examine his reasons and perhaps to advance the analysis of a crucial problem beyond where he left off. Siegler and I will disagree, not necessarily about his decision not to intervene but about the underlying assumptions that shaped his interactions with Mr. D. I intend to highlight the significance of the underlying assumptions that guided his conduct rather than the decision itself, for unless these assumptions are better clarified, the climate of physician-patient decision making will not improve significantly.

Some years ago Mark Siegler, while an attending physician at a teaching hospital, met Mr. D., a previously healthy sixty-six year old black man who had come to the emergency room suffering from an acute febrile illness of three days' duration. The patient appeared to be critically ill and was admitted to one of the medical wards. A chest X-ray demonstrated a generalized pneumonia and he was treated aggressively with three antibiotics. The next day Mr. D.'s condition worsened and his physicians concluded that two uncomfortable but relatively routine diagnostic procedures might establish the cause of his illness: a bronchial brushing to obtain a small sample of lung tissue and a bone marrow examination. The patient refused permission for these tests, and when his physicians repeatedly attempted to explain their necessity, Mr. D. "became angry and agitated by this prolonged pressure, and subsequently began refusing even routine blood tests and X-rays" ([14], p. 12).

A psychiatric consultant found him competent and concluded "that Mr. D. understood the severity of his illness ... [and] that he was still making a rational choice in refusing the tests" ([14], p. 12). The patient's condition deteriorated and twenty-four hours later he appeared to be near death. He then refused to be placed on a respirator and Siegler was unable during two forty-five minute interviews at the bedside to get Mr. D. to change his mind. At the same time everything that Mr. D. had said convinced Siegler

> that he understood of his situation. For example, when I told him he was dying. he replied: "Everyone has to die. If I die now, I am ready." When I asked him if he came to the hospital to be helped, he stated: "I want to be helped. I want you to treat me with whatever medicine you think I need. I don't want any more tests and I don't want the breathing machine."

I gradually became convinced that despite the severity of his illness and his high fever, he was making a conscious, rational decision to selectively refuse a particular kind of treatment. In view of the frankness of our discussion, I then asked him, whether he would want us to resuscitate him if he had a cardio-respiratory arrest. He turned away and said: "We've been through this before; now leave me alone."

Mr. D. soon became semi-conscious and had a cardio-respiratory arrest. Despite the objections of the houseofficers, I did not attempt to resuscitate him, and he died. ([14], p. 12)

Siegler identified six factors that influenced his decision "to support [Mr. D.'s] choices." They were based on premises of "clinical ethics" that "take into account and reflect the extraordinary complexity of the medical model." At the same time his ultimate decision was significantly affected also by his "belief in the rights of individuals to determine their own destinies" ([14], p. 14), considerations not generally encompassed by the "medical model." Before analyzing Siegler's criteria, let me repeat once again that I do not wish to quarrel with his judgment of not subjecting Mr. D. to diagnostic tests, not placing him on a respirator, or not attempting resuscitation. Indeed, in the light of the tragic choices that confronted him and of the opposing sentiments of the house staff, I admire his courage to stick to the decision that made sense to him. If some will consider his judgment flawed, a contrary judgment may have proven equally flawed. Whether I would have behaved differently I cannot say. Much depends here on whether my way of thinking and proceeding would have provided additional information about Mr. D.'s refusal. I am only certain of one thing: I too *might* have ultimately acquiesced in or supported Mr. D.'s choice.

Before proceeding, I must also draw attention to the fact that Siegler's account of his and the psychiatrist's interviews are devoid of data as to why Mr. D. was so adamant in his refusal. I shall assume that Siegler and the psychiatrist were unable to elicit any information from Mr. D. about his reasons for refusing the tests. That assumption is crucial for what follows, for Siegler may have data on that issue which, if he had presented them in this article, would have led me to a similar decision more easily. But, as I have already suggested, I am less interested in the decision than in the underlying assumptions.

These are Siegler's six factors: (1) "*The patient's ability to make (rational) choices about his care.*" That determination depends on a physician's assessment of whether a patient has "retained sufficient intellect and rationality to make choices." It constitutes a "subjective clinical judgment" ([14], p. 13). (2) "*The nature of the person making the choice.* That determination depends on one's knowledge about the patient's "personality, character, ideas and beliefs," in short on an assessment of whether the patient is "acting autonomously - that is, with authenticity and independence" ([14], p. 13). The family here can provide useful information. In Mr. D.'s case family members could not be located and Siegler learned little about his personality, character and beliefs. In time-limited situations this is frequently the case and thus, according to Siegler, "clinical judgment" plays an important role in assessing this factor. (3) "*Age.*" Siegler postulates a direct relationship between increasing age and non-intervention. "The closer a patient gets to a 'normal' life span, the more he has lived, and the more ready I am to let 'nature take its course' " ([14], p. 14). (4) "*Nature of the illness.*" Good prognosis, probability of success, according to Siegler, suggest intervention, while uncertainty of diagnosis or uncertainty of prognosis, particularly in the presence of progressive disease, suggest respect for patients' wishes not to be treated. (5) "*The attitudes and values of the physician responsible for the decision.*" Siegler discusses under this rubric physicians' values and their impact on decisions. He acknowledges that physicians' values differ and that his "belief in the rights of individuals to determine their own destinies" ([14], p. 14) affected his decision not to intervene in the case of Mr. D. (6) "*The clinical setting.*" Siegler notes here the influence of the institution, for example, the complex relations that exist in a teaching hospital between attending physicians and house staff, on decision making.

In a subsequent article, co-authored with Ann Dudley Goldblatt, Siegler omitted, without explanation, the fifth and sixth factors from consideration [15]. I believe I understand why he did so, for they, unlike the first four factors, do not address the patient's condition or his capacities to participate in the decision-making process but focus instead on physicians' attitudes and the setting in which they work. Although these factors require separate consideration, they must not be eliminated, since they crucially affect the decision-making process.

Indeed, they are initially more significant than the capacities and incapacities that patients bring to the physician-patient encounter. Physicians' incapacities to reason have not received the same attention as patients' incapacities. Doctors must become more sensitive to the impact on their conduct of personal, professional, and institutional value orientations, be they their attitudes toward death, their constant quest to defeat the grim reaper, the importance they ascribe to age, their attitudes toward acute and chronic, reversible and irreversible illness, their views about the patient as a worthy or unworthy partner in the decision-making process, the deference they give to colleagues, house staff and the institution itself, and much more. Unless physicians consider and sort out these matters prior to their first encounter with a patient, the decision-making process is fatally flawed *ab initio*. Thus, I would reverse the order of Siegler's factors, and combine factors five and six, and put them at the head of the list. The emphasis must be shifted from the capacities and incapacities of patients to those that physicians bring to their encounter with patients. Doctors initiate the dialogue and how it begins powerfully influences what follows.

I would eliminate age as a crucial separate factor. It does not deserve to be so highlighted and surely not as an "objective" criterion. It belongs under the rubric of "attitudes and values of the physician." Considerations of age, unless supported by clear and convincing evidence from the patient's mouth, can deceive, and dangerously so. I am almost Mr. D.'s age. I love life and I would not wish my doctor to "let nature take its course," if something could be done for me.

It might be objected that, had I felt this way, I would not have behaved as Mr. D. did. Yet, this is only half the story. If age considerations are viewed as an objective criterion and not as a value preference, then, in situations of initial refusal, when the reasons for refusal are still murky, the factor of age may influence the physician to desist from trying as hard as he would otherwise to learn why an aged patient seems so adamant in his refusal. This is dangerous.

Moreover, considerations of age in situations of both outright refusal and patients' willingness to consider some treatment alternatives but not others, may subtly influence the extent of communication. With older patients some physicians may be led to think that it matters less than with younger patients whether the medically indicated procedure is fully understood and, in turn, chosen. This is even more dangerous. Older patients may cherish their remaining years and the quality of their lives more than some younger patients do. Consideration of age constitutes a value judgment or preference, no more, no less.

Siegler's fourth factor, "the nature of the illness," encompasses two features that must be distinguished: physicians' *and* patients' attitudes toward the nature of the illness. The two do not necessarily coincide. Illness, its readiness or refractoriness to yield to diagnosis, its certain, uncertain, grave or hopeful prognosis, affects physicians in powerful ways. Doctors must sort out first how such considerations influence their conduct before trying to ascertain patients' attitudes toward their illness.

Otherwise, the danger is great that initial refusals will be too readily "honored," yet influenced more by physicians' attitudes about the futility of treatment than patients' willingness to prolong life a bit longer, even at considerable personal cost.

Shifting the focus from the patient to the nature of his illness invites confounding a professional judgment of its severity with a personal judgment of what severity may mean to the patient. And if doctors, on the basis of prognosis, support refusal all too readily, they may give patients the impression, however much unintended, that the latters' lives are not worth saving, that doing so is too time-consuming, too expensive, too inconsiderate of other patients' needs. Such impressions may reinforce patients' declarations of refusal, even though deep-down they yearn to be convinced otherwise. My colleague Robert Burt has eloquently spoken to this problem while analyzing another encounter between a physician and a patient who seemingly preferred death over continued treatment.

> The critical ambiguity that I see in [Mr. G.'s] conversation with Dr. White goes rather to Mr. G's conception of himself as a choice maker; that is, it is not clear whether he sees himself as separate from others in exercising choice regarding his future or whether he chooses death because he believes others want that result for him and he feels incapable of extricating himself from their choice making for him. Either perspective could lead him to the deepest despair: his affliction itself could rob life of all possible meaning for him; his belief that others found him repellent, contemptible, and wished him dead could do the same. [Thus it is] critical to establish which of these two perspectives led him to this choice. Was he, in other words, implementing his own choice because he saw himself as free to do so, or was he implementing the choice of others because he saw himself as inextricably bound to their choice for him? ([3], p. 11)

The factor of "nature of illness" must not be taken as an objective criterion for either respecting or disregarding patients' refusals. At least, it must be recognized that too much can be made of it, particularly by the physician; and that it invites premature closure in ascertaining what the patient wishes to make of his illness.

This leaves two of Siegler's factors that I intend to combine: "the patient's abilities to make choices about his care" and "the nature of the person making the choice." They are indeed crucial. Siegler repeatedly emphasized, as is so customary in medical practice, *his* "assessment" of Mr. D.'s rationality, authenticity, intelligence, and independence. Such an assessment was also made by the psychiatric consultant. Yet these unilateral assessments that implicitly, if not explicitly, stress the patient's irrationality and not the physician's can readily compromise *mutual* exploration from its very beginning. They tend to fuel a patient's resentment over not being taken seriously, over having to establish his credentials as a competent person, however much unintended by the physician, as Siegler's caring concern demonstrated.

Required instead is a bilateral conversation between doctor and patient that explores *their* expectations of one another, that identifies *their* misconceptions, *their* confusions and, most importantly, that seeks to clarify why *they* wish different things from one another. All this must be done in the spirit not of assessing, evaluating, or judging anything but of better understanding one another.

I would reformulate, and reduce to three, the factors that Siegler urges physicians to consider in situations of patients' refusal. For now I shall list them with only the briefest of commentary. In the next and final section of this essay I shall attempt to justify my priorities and formulation:

(1) *The attitudes and values that physicians bring to the decision-making process.* This factor must precede all others. Unless physicians are clearer about their attitudes and values, they can neither initiate conversation with patients nor know what to disclose to patients. Doctors need to reflect more about their value preferences, how these values shape their attitudes toward age, treatability, prognosis, the primacy of life, death with dignity, and patients as partners in decision making. Above all, they must reflect more about their own vulnerabilities to idiosyncratic, unconscious, and irrational thought and action.

(2) *The attitudes and values that patients bring to the decision-making process.* Doctors must learn to appreciate that rarely, as proved so true for Mr. D., is patients' "competence" in question but rather patients' vulnerabilities to unconscious and irrational fears and expectations, to confusion and misconceptions, to investing physicians with omnipotent or destructive powers (depending in part on whether they viewed their earliest caretakers as caring or neglecting), to submission or rebellion because they feel so utterly disregarded and not taken into account. Moreover, doctors must learn how patients' values about illness, disability and death affect their expectations of doctors and of what they wish doctors to do for them.

(3) *The obligation for mutual conversation in order to clarify the expectations that both physicians and patients have of one another.* Both

parties must appreciate that they have much to learn from one another, that they can take little for granted, that physician and patient influence one another in decisive ways, and that both must entertain the possibility of reconsidering their initial judgments of how best to proceed. Who will prevail in situations similar to Mr. D.'s must remain an open question until the moment comes when a decision has to be made, yet with the presumption, although on rare occasions a rebuttable one, that the patients' ultimate choices are more decisive than the doctors'.

My criteria impose an obligation on both physicians and patients to engage in conversation so that they will clarify with one another their differing views. Perhaps in the case of Mr. D. it would then have emerged that his refusal to undergo diagnostic tests was influenced by the fact that "ten years earlier he had left [another] hospital 'against medical advice' after first refusing to have a bone marrow examination" ([14], p. 13). This information that became available only after his death is of more than "anecdotal interest." It graphically illustrates how unexamined prior experiences – although it may not have been true for Mr. D. – can condemn one to death.

III. WHEN AND WHY SHOULD PATIENTS' WISHES AND CHOICES NOT BE RESPECTED?

Siegler's decision to defer to Mr. D.'s wishes was decisively influenced by his "belief in the rights of individuals to determine their own destinies" ([14], p. 14), a belief that comports with a fundamental Anglo-American jurisprudential principle - the right to individual self-determination or, as it also has been called, the right to autonomy. Before directly addressing these principles and their significance in situations in which patients refuse to comply with doctors' recommendations, I must pause and first talk more generally about self-determination, autonomy and a number of related concepts. In my recent book, I have made an attempt to re-examine these concepts so that they comport better with assumptions about the psychological nature of human beings [9].

Let me begin with a few broad definitions that I shall refine as I go along. The right to self-determination has generally been defined as the right of individuals to make their own decisions without interference from others. Under this broad definition, once a person has made a decision, he has a right to proceed regardless of how the individual reached the decision. However, in the context of physician-patient decision making, I mean, as I shall soon make clear, something more focused than this broad definition.

Autonomy has often been used interchangeably with self-determination, but I shall employ the concept of autonomy, or "psychological autonomy" as I shall call it, to denote solely the capacities of persons to exercise the right to self-determination. In my scheme, psychological autonomy addresses persons' capacities to reflect about contemplated choices and to make choices. The extent and limits of such capacities of course vary from individual to individual, but I am not concerned here about such individual variations. I wish to focus instead on those capacities for self-determination that depend on one's views of the psychological nature of human beings. In essence psychological autonomy is a concept that "informs" the principle of a right to self-determination by explaining, refining, and pointing up human capacities and incapacities for the exercise of such a right.

I have given autonomy's root meaning - *autos* and *nomos* = self law – an unaccustomed construction, for traditionally it has emphasized different areas of rights, be they political, legal or moral rights [2, 17]. I wish to emphasize, however, the psychological capacities that underlie rights, including the right to self-determination. Indeed, underlying, although generally unidentified, assumptions about human psychology have shaped decisively the views on all political, legal, and moral rights to which persons are supposedly entitled. I retain the term autonomy to call attention to the fact that traditional definitions of autonomy, and of rights as well, contain a great many psychological assumptions that have been given insufficient consideration in understanding the complexities of decision making between physicians and patients.

Let me put this another way and at the same time introduce a number of other distinctions. Self-determination contains, as my discussion on psychological autonomy will soon try to clarify, two intertwined, though separable, ideas: One looks to conduct in relation to the external world, at conduct in relation to action. This external component of self-determination I call *choice.* It has also been spoken of as freedom of action. The other looks at conduct in relation to the internal world, at conduct in relation to thinking about one's choices prior to action. This internal component of self-determination, I call *reflection* or *thinking about choices.*

A number of considerations have led me to introduce distinctions between internal reflection and external choice, as well as to highlight the importance of psychological autonomy. I shall begin with the external-internal distinction.

(1) Separating self-determination into its internal and external components permits posing two sets of fundamentally different questions: One "external" question: "To what extent should an individual's *choices* be respected," and two "internal" questions: "To what extent should an individual's *thinking about choices* be respected? By what means, if any, can and should a person's capacity for reflection be enhanced through conversation? The last question is important if I am correct in assuming that human beings' psychological capacities for self-determination are limited, yet subject to enhancement by conversation. Any answers one might wish to give to these questions will be affected by one's views on psychological autonomy but, as I intend to argue, they do not necessarily lead to the conclusion that the same respect must be accorded to both choice and to thinking about choices.

(2) In all human encounters, including those between physicians and patients, assumptions about human beings' psychological capacities to think and to act influence attitudes toward choice and deliberations. (I shall use deliberation to denote both self-reflection and reflection with others.) The only psychological assumptions that the medical profession has acknowledged and endorsed are that patients, by virtue of illness and ignorance about medical matters, are incapable of making decisions on their behalf. The external-internal distinction draws attention to a crucial question that challenges this assumption: If physicians were to provide patients with a meaningful opportunity for conversation, could whatever incapacities to reasoning illness engenders be moderated sufficiently so that patients' choices could be accorded greater respect? It is likely that the answer would be affirmative for a significant number of patients, particularly those who do not suffer from acute illnesses. If physicians were to pay greater attention to conversation - to patients' capacities to reflect about choices - it could change their traditional attitudes toward patients' capacities to make their own decisions and, in turn, could radically transform the current unsatisfactory state of physician-patient decision making.

The importance one wishes to assign to the internal and external components of self-determination, however, will also be shaped by how one weighs political, legal, and moral value preferences – such as privacy, beneficence, loyalty and freedom. Looked at from this perspective, the external component also encompasses the choices individuals must be allowed to exercise, to uphold these values (e.g., respect for liberty and freedom of action), and the internal component encompasses the extent of reflection and conversation that individuals should be obligated to engage in prior to acting, to respect these values. For example, respect for the great importance physicians place on beneficence and loyalty to their patients may suggest that physicians have a right and need to be informed why their patients do not choose to follow a proposed course of action so that they can be reasonably certain that patients have understood their doctors' recommendations.

(3) The internal-external distinction also permits separate consideration of the relevance of assumptions about human psychological functioning to reflection, on the one hand, and to choice, on the other. Discussions of patients' right to self-determination largely have focused on choice, with proponents of self-determination asserting that the choices of all but incompetent persons must be honored and that the introduction of any psychological assumptions is irrelevant or even treacherous. For the most part, I share these views. Even though choices are influenced by psychological considerations, it is one thing to appreciate that fact and quite another to interfere with choice on the basis of speculations, or even evidence, about underlying psychological reasons that seemingly led a patient to make the "wrong" choice. Such psychological explanations can be found too readily and exploited too easily for purposes of over-reaching and coercion. Thus, the danger of interfering with patients' choices on psychological grounds are too great and too difficult to control. Short of overwhelming evidence of incompetence, choices deserve to be honored.

The process of thinking about choices, however, deserves to be treated differently. Here psychological assumptions suggest that physicians and patients are under an obligation to reflect and to converse. Ignorance, misconceptions, exaggerated fears, and magical hopes about diagnostic tests and therapeutic interventions, as well as about what physicians and patients want and are able to do for one another, can decisively influence thinking about choices. The danger is great that patients' and doctors' choices will be distorted by such internally and externally engendered mistaken ideas. Thus, conversation will have to be more extensive and more searching if one believes that such distortions affect one's thinking about choices and that they deserve to be clarified to the extent possible. On the other hand, such conversation can be more limited if one holds to the belief that thought and action are based largely on a reasoned awareness of the motivations and reality factors that influence physicians' and patients' conduct.

The process of thinking about choices, to return once again to what I have already asserted, will be significantly altered if it were appreciated that physicians' capacities and incapacities for reflection and choices are as vulnerable to distortion as those of their patients. The contrary assumption that doctors' contributions to any conversation, in contrast to patients' contributions, are influenced largely by rational (personal and professional) considerations, has made both self-reflection by physicians and searching conversation between them and their patients from which *both* can learn seemingly irrelevant. If the influence of idiosyncratic psychological determinants on physicians' communications – idiosyncratic because they are so rarely exposed to conscious scrutiny – were recognized, it would affect decisively the parties' perceptions of one another and, in turn, their mutual deliberations.

My views on psychological autonomy, to which I now turn, may further clarify why I attribute such importance to the distinctions I have introduced. The psychological assumptions that inform my conception of psychological autonomy emphasize the admixture of consciousness and unconsciousness, rationality and irrationality in all human thought and conduct. It is never one or the other, though the mix may differ in different persons, and for the same person under the impact of different external conditions, The relative impact of the unconscious and the irrational is also affected by the quality of the conversation between persons.

Rationality refers to the impact on thought and action of consciousness, reality needs, time perspectives, varied and subtly blended emotions, realistic expectations, necessary postponements of gratification, reflective thought prior to action, and regard for facts. Irrationality refers to the impact on thought and action of unconscious impulses and ideations, fantasies, timelessness, concreteness, unmodulated emotions, confusion of past and present realities, unattainable and infantile conceptions, and disregard of facts ([13], p. 123).

My definition of autonomy, in contrast to those of others, therefore, is not solely "linked ... to those higher forms of consciousness which are distinctive of human potential" ([6], p. 74). I reject its artificial restriction to conduct regulated primarily by reason. Since reason and unreason always act in concert, both are facets of psychological autonomy. These views have several major implications:

(1) Thought and action can never be brought fully under the domination of consciousness and rationality. Physicians and patients, however, can become more aware of the ever-present and pervasive impact of unconsciousness and irrationality on thought and action. An appreciation of the limits of rationality would sensitize physicians and patients to the need for subjecting their thoughts and contemplated actions to prior reflection, both alone and with one another. It would also sensitize them to the inevitable persistence, even after conversation, of lingering, nagging, and quite healthy doubts about the continuing influence of unconscious and irrational determinants on their conduct. Such an awareness, which unites physicians and patients in common vulnerabilities, can only make them more careful about apodictic judgments and more respectful of their respective wishes, particularly when they are in conflict with one another.

(2) The simultaneous domination of thought and action by rational and irrational, conscious and unconscious determinants can be shifted in the direction of greater rationality and consciousness. At a minimum, ill-considered and mistaken ideas can be clarified, uncertainties specified, fears dispelled, and the confusions of past experiences that have no relevance to the present situation untangled. Self-reflection and reflection with others can aid this process immeasurably. Thus, physicians and patients need to engage in conversation to clarify their mutual expectations to the extent possible. Most importantly, physicians must assume the primary obligation of facilitating such conversation.

Indeed, I shall take a further step and postulate a duty to reflection that cannot easily be waived. Asserting such a duty sounds strange. We are accustomed to recognizing a right to choice as an aspect of the right to self-determination, but a duty to reflection, as a component part of the right to self-determination, is quite another matter. Yet, if my views on autonomy have merit, then respect for the right to self-determination requires also a respectful awareness of human beings' proclivities to exercise this right in both rational and irrational ways. To prevent, to the extent possible, any ill-considered irrational influences on the choices made, yet ultimately honor whatever choices are made, demands at least that doctors become obligated to facilitate patients' opportunities for reflection.

Patients, in turn are obligated to participate in the process of thinking about choices. In arguing that both parties make every effort to facilitate reflection, I seek to pay respect to the principle of psychological autonomy as I have defined it; that is, to both the capacities and incapacities of human beings for rational choice.

In my view, the right to self-determination about ultimate choices cannot be properly exercised without first attending to the processes of self-reflection and reflection with others. This holds true for patients as well as for physicians. Contrary views have insufficiently considered not only human proclivities for unconscious and irrational decision making but also, and more importantly, the possibilities of bringing some of these determinants to greater awareness. Such views on autonomy and self-determination do not pay respect to "self-defined" persons; instead, such views inhibit opportunities for women and men to become clearer about how they may wish to define themselves, abandoning them in the process to malignant fate. In the context of physician-patient decision making, it must be recognized that illness - including the fears and hopes it engenders, the ignorance in which it is embedded, the realistic and unrealistic expectations it mobilizes - can contribute to tilting the balance further toward irrationality and choices that, on reflection, both might wish to reconsider. In short, I seek to justify the duty of reflection on the grounds of human beings' capacities to take their unconscious and irrationality more fully into account.

I am not suggesting, however, that the deliberations between physicians and patients be converted into an exploration of the psychological roots of patients' and physicians' motivations and expectations. This is neither warranted nor possible. I have in mind only a bona fide attempt by physicians and patients to explain what they wish from and can do for and with one another, and to clarify, to the extent possible, any misconceptions they may have of each others' wishes and expectations. In the end, irreconcilable differences may persist. If they then realize that they must part company, at least they will do so with a greater appreciation of their respective positions.

(3) The high value I place on the requirement for conversation to protect autonomy can create inevitable conflicts with the right to privacy – the right to keep one's thoughts and feelings to oneself. Refusals to converse, however, may totally obscure both patients' and doctors'

understanding of how both arrived at their decision. This is particularly true when patients either decline a needed medical intervention or accept it unquestioningly. The capacity for exercising psychological autonomy optimally becomes compromised when refusals or acceptances are heeded without question. Here the principle of privacy must bend to psychological autonomy. This may turn out to be a rare Hobson's choice, for I expect that most patients, if invited by their physicians, will welcome conversation.

While the imposition of an obligation to converse rules out an absolute right to have one's initial choice, including the right not to converse, honored, it seeks to safeguard ultimate, considered choices that are more self-reflective. The process of thinking about choice must, of course, not be exploited in order to obtain a spurious consent. Since human beings are not only actors but reactors as well, exploitation of the process of thinking about choices is an ever-present danger that only a commitment to mutuality of decision making can guard against. The current low status of mutual decision making in medical practice may be more the doctors' than the patients' fault, and thus it is particularly the physicians' contributions that my recommendations seek to reverse.

(4) The posited obligation to converse introduces an element of paternalism into my prescription. Yet, it must also be recognized that my views about psychological autonomy and its accompanying obligations to reflect and converse seek to avoid paternalism and to strengthen patient self-determination. The obligations that I advocate are imposed on *both* parties; they do not ask for one party to submit to the other; they are grounded in mutuality; and they are dictated by a respect for human psychological functioning in the specific context of physician-patient decision making: patients' wishes to be an adult *and* a child, and doctors' wishes to treat patients more as children *than* as adults.

Awareness of the complexities of psychological autonomy requires that both parties pay caring attention to their capacities and incapacities for self-determination by supporting and enhancing their real, though precarious, endowment for reflective thought. In conversation with their doctors, patients may uncover mistaken notions about their disease and its treatment that they have held for a long time or have recently acquired through misunderstanding the import of their doctors' recommendations. In conversation with their patients, physicians may uncover that unconscious preferences and biases compelled patients to yield to doctors' recommendations even though consciously patients had intended otherwise. Without conversation, individual self-determination can become compromised by condemning physicians and patients to the isolation of solitary decision making, which can only contribute to abandoning patients prematurely to an ill-considered fate.

It has been said that remonstrances against paternalism, while logically not inconsistent with concern for others, may nevertheless diminish one's concerns for others. Karl Marx said it more strongly: "The right of man to liberty is based not on an association of man with man, but on the separation of man from man. It is ... the right of the restricted individual, withdrawn into himself" ([11], pp. 162-163). While there is considerable truth in these statements, it must be remembered that except in the rarest of circumstances, the ultimate decision belongs to the patient who has to live with the decision. One can only try to help a patient to make *his* or *her* best choice. Both of these considerations loom large in my distinction between reflection and choice.

Before returning to Mr. D. and the problems posed by his refusal to undergo diagnostic tests, let me emphasize again my conviction that generally patients' choices deserve the utmost respect and irrespective of underlying psychological incapacities, short of legal incompetence. Human beings' unconcious and irrationality are part of their essence. If unaffected by conversation, their choices generally deserve respect because coercion in the name of so-called psychological health can easily run wild. Thus, in the evaluation of choice - the external component of self-determination - considerations of human beings' psychological autonomy should not be decisive. Although this means bowing at times to "foolish" choices, they must be honored nevertheless, perhaps less out of respect for the freedom of the individual patient involved than out of respect for the freedom of all patients whose capacities to think about their choices will be fatally impaired if the threat of having their choices vetoed whenever they appear foolish hangs forever over their heads. I am sufficiently optimistic that if the process of thinking about choices - the internal component of self-determination - is attended to with greater care, the problem of a stand-off between physicians and patients will be a rare event. In making such a prediction I do not wish to suggest that patients will always follow doctors' recommendations but rather that in the face of disagreement both will learn to appreciate that their irreconcilable differences have merit and deserve respect.

Yet, there are times when a patient's choice should not be honored. Mr. D. may or may not be an example in point. Since disregard of patients' wishes is an awesome judgment to make, it must be carefully circumscribed and justified. I would entertain the possibility of doing so in situations (1) when the consequences of non-intervention pose grave risks to a patient's immediate physical condition *and* (2) when the process of thinking about choices is so seriously impaired that neither physician nor patient seem to know what one or both wish to convey to the other.

Thus, I would limit interferences with patients' choices to situations of illnesses that might terminate in death or lead to irreversible complications and for which diagnostic and therapeutic interventions seem to be available that might reverse or control the disease process. That death and irreversible damage deserve concern I need not defend. They are, however, only *necessary,* but not *decisive,* conditions for intervention.

Interference with patients' choices must also be justified on grounds of the flawed nature of the process of thinking about choices. I shall not repeat what I have said earlier about the importance I attribute to self-reflection and reflection with one another. I only wish to emphasize once again that *both* physicians and patients are obligated to engage in a searching conversation to explain their respective positions to one another. *Neither physician nor patient can easily repudiate that requirement,* at least not the physician, and the patient only after he has made a good case for doing so.

Earlier I have attempted to justify an obligation to converse on the basis of my views on psychological autonomy. I now wish to advance two other justifications: one grounded also in human psychology and the other in the special nature of the physician-patient relationship. First, reflection with others, as psychoanalysis has taught us, facilitates awareness of the determinants that affect thought and action to a greater extent than can be accomplished by solitary self-reflection. Second, in medical decision making an initial fundamental inequality of knowledge exists between the parties, with the physician knowing much more about medicine and the patient much more about himself. These inequalities can only be bridged by their sharing their knowledge so that both understand the impact of their respective expertise on the medical decision to be made. Without such sharing, choices will only be spurious, if not meaningless.

Some patients, to be sure, may wish to abdicate the responsibility for joint decision making and delegate it solely to their physicians. Yet, even if doctors are agreeable to such an arrangement, they must first insist that patients give a satisfactory account of their reasons for doing so. At least for some time to come, in the light of the traditional climate in which patients are expected and expect to follow doctors' orders, inquiry may establish that patients abdicate decision-making responsibility for reasons that doctors should question and put to rest; for example, patients may, on inquiry, reveal their concerns about imposing on physicians' time, their fears about annoying doctors with their questions, their embarrassment about not being able to ask the right questions, without assistance from their doctors. These and other misconceptions need to be dispelled. If then other good reasons emerge for delegating decision-making authority to physicians, they should be accorded the respect they deserve.

In situations of illness that threaten life or can cause irreversible damage, patients, however, cannot both refuse intervention and waive the requirement of conversation. Had I encountered Mr. D., I would have told him that I was puzzled by his refusal to undergo the proposed diagnostic tests. I would have expressed to him my concern and confusion over my lack of understanding of what had led him to his decision as well as my concern and fear of perhaps not having adequately conveyed to him why I thought these tests were so essential to his well-being.

I would have impressed on him the necessity of our talking together. Indeed, I would have insisted on our talking together as long as time permitted in order to clarify our respective positions. I would have promised him that I had every intention ultimately to respect his wishes, but that I could not make an absolute promise to do so, for it could turn out that the acuteness and seriousness of his condition might require intervention prior to our having made ourselves understood to one another. I would have added that I expected this to be an unlikely outcome, but that it could happen.

If in the midst of our talking together, Mr. D. had "turned away" and bid me to "leave [him] alone" ([14], p. 12), I would not have left his bedside. Such conduct would have been tantamount to acting as if I had not been there in the first place, as if I could be totally obliterated before Mr. D. had succeeded in obliterating himself [3]. I would have felt impotent and experienced Mr. D. as all too powerful. And I would have recalled what Burt had written: that such depictions of myself and him might have enraged me and made me turn away out of an unconscious wish to hurt him [3]. At the same time, I might also have overlooked that the patient, appearances notwithstanding, was struggling with feelings of impotence out of the intense stress he was feeling from the incapacitating experience of his illness. Thus, I would have stayed with him and renewed my invitation to continue talking. Had he declined the invitation, I would eventually have been forced to tell him that I might very well order the tests, place him on a respirator, and resuscitate him if he refused to talk with me. "There is too much that we both do not understand," I would have added, "and you must not hide behind silence."

Who knows how my insistence to talk would have affected Mr. D? It might have pierced his isolation and uncommunicativeness, and led him to express his concern and confusion over the proposed tests. Perhaps he viewed them as being of experimental rather than diagnostic necessity. After all, some patients are convinced that doctors at university hospitals like to conduct experiments on patients ([5], p. 125). In this connection, let me note again that Siegler learned only after Mr. D.'s death that "ten years earlier [Mr. D.] had left a hospital 'against medical advice' after first refusing to have a bone marrow examination" ([14], p. 13). Since in the intervening years he had not been any worse off for his refusal, he might have wondered whether this test was not equally unnecessary now. Had he told me about this experience, perhaps I could have impressed on him that the two situations were not necessarily comparable. From there, we could have gone on to talk about his mistrust of doctors and the uncertainties of medicine. Who knows what else we might then have explored?

Moreover, Mr. D. clearly wished to be treated and *only* refused the performance of diagnostic tests. Something incomprehensible was at work here that begged for clarification. And if it could not be clarified, how is one to judge the state of his autonomy? It seems that a conflict existed between his "autonomous" wish to be cured and his "autonomous" refusal to undergo diagnostic tests. Which deserved respect?

If on the basis of my nagging and unanswered questions I had intervened, I would not have done so because I thought his decision unwise, foolish or whatever, but because I had no idea why he had decided what he did. I would have felt confused and been uncertain whether he was confused as well.

I appreciate the problems and dangers of so conducting myself. I can only acknowledge my discomfort and at the same time observe that remaining dumb and deaf would have made me even more uncomfortable. While I am committed to the principle of respect for patients' choices, I also believe that no principle can rule absolute. The principle of self-determination, its external component, is no exception. At the same time, any exception must be narrowly circumscribed and justified, and more carefully than I could do in this essay. Here I only wanted to establish the principle of at times not respecting patients' choices out of a respectful awareness of its companion concept, psychological autonomy, and to define its general limits.

Let me, however, make two additional comments about justifications for a rare exception. Physicians also have needs that deserve respect. In situations like Mr. D.'s, their strong ethical commitment to caring for patients can impose intolerable burdens on them. In these instances, doctors may never know whether they have explained themselves satisfactorily to their patients. Doctors may then doubt whether they have taken the necessary time or made the necessary effort to make themselves understood. Such doubts can lead to nagging guilt feelings over having failed in their professional obligations. Assuaging guilt-inducing doubt that may haunt physicians for a lifetime is another reason for my insistence on conversation.

Equally important is another consideration, and Mr. D.'s case highlights the problem. If all final authority is vested in patients, the danger is great that in situations of a total refusal to give an account of one's reasons or of an unwillingness to explore one's possible confusion – when the need for conversation is the greatest – doctors will wittingly and unwittingly give up on conversation and patients prematurely because they have been stripped of all power to stop even patients' most inexplicable self-destructive course. To protect doctors and, in turn, patients from such pernicious consequences supports the creation of a *rare* exception to the rule that doctors otherwise must obey: In case of disagreement, doctors and patients should either go their separate ways, or agree to provide and to receive care within the limits imposed by the patient.

At a minimum, establishing a rationale for instances in which refusals should not be honored might ameliorate the danger of manipulating consent engendered by fears that an open confrontation of differences must always result in submitting to patients' choices. Moreover, a careful scrutiny of cases in which patients' choices were not honored will instruct us about why doctors have to proceed in this fashion, what could have been done differently, and why. In this day and age, coerced, manipulated, and consented-to interventions are so hopelessly intertwined that any exploration of why physicians and patients disagree is well nigh impossible.

If what I set forth makes readers uncomfortable too, they will only have identified with my discomfort. But the practice of medicine is the practice of an excruciatingly demanding profession, not only because of the inherent limits in its capacities to cure but also because of the inherent limits in human capacities to interact with one another reasonably, sensibly, rationally, and caringly. These limitations take on special significance in confronting the dilemmas that are the subject of this essay. The central dilemma, as the writers of Genesis knew so well, is the responsibility of one human being to another in the face of unnecessary death. After Adam and Eve had eaten from the Tree of Knowledge and had tasted the joys and sorrows of human freedom, the responsibility for another person's life and death was the first issue the writers of Genesis addressed.

When, after the slaying of Abel, God confronted Cain and asked, "Where is Abel, your brother?" Cain answered, "I do not know; am I my brother's keeper?" [7]. He did not leave it, though he could have, by merely saying "I do not know." He added a question: "Am I my brother's keeper?" And it has remained a question ever since: "What does it mean to be my brother's keeper?"

In Hebrew *shamar* (to keep), the verb of *shomer* (keeper) is one of those key words which has many meanings: "to keep, to retain, to protect, to safeguard, to preserve, to guard, to nourish, to observe, to celebrate and to wait." My inquiries into a radically different kind of dialogue between physicians and patients, into the respect that must be accorded to patients' choices on the one hand, and to their disregard when death or irreversible injury can be avoided, on the other, are also attempts to give meaning to the extent and limits of physicians' obligations toward patients as their keepers. Should physicians be patients' guardians (a meaning of "to retain"), respect their needs (a meaning of "to nourish"), respect their individuality (a meaning of "to celebrate"), delay until they are ready to give us the authority to proceed (a meaning of "to wait"), or intervene despite their wishes (a meaning of "to preserve")? It is an inquiry worth pursuing, for it will protect, safeguard, preserve, celebrate, nourish, and enrich physicians and their patients, and the art and science of medicine as well.

Yale Law School
New Haven, Connecticut

BIBLIOGRAPHY

[1] American Medical Association: 1981, *Current Opinions of the Judicial Council,* American Medical Association, Chicago.
[2] Beauchamp, T. and Childress, J.: 1979, *Principles of Biomedical Ethics,* Oxford University Press, New York, pp. 56-57.
[3] Burt, R. A.: 1979, *Taking Care of Strangers,* Free Press, Macmillan Publishing Co., New York.
[4] Cahn, E.: 1961, 'The Lawyer as Scientist and Scoundrel - Reflections on Francis Bacon's Quadricentennial', *New York University Law Review* **36**, 11.
[5] Duff, R. and Hollingshead, A.: 1968, *Sickness and Society,* Harper & Row, New York.
[6] Dworkin, G.: 1982, 'Autonomy and Informed Consent', in President's Commission for the Study of Ethical Problems in Medicine and Biomedical and Behavioral Research, *Making Health Care Decisions,* Vol. 3, Washington.
[7] Genesis, Chapter 4, 9.
[8] Kant, I.: 1964, *Groundwork of the Metaphysic of Morals,* H. Paton (trans.), Harper & Row, New York.
[9] Katz, J.: 1984, *The Silent World of Doctor and Patient,* Free Press, Macmillan Publishing Co., New York.
[10] Ladd, J.: 1978, 'Legalism and Medical Ethics', in J. W. Davis and C. B. Hoffmaster (eds.), *Biomedical Ethics,* Humana Press, New York, pp. 1-35.
[11] Marx, K.: 1974, 'On the Jewish Question', in K. Marx and F. Engels, *Collected Works,* International Publishers, Moscow.
[12] *Natanson v. Kline,* 350 p. 2d 1093 (Kan. 1960).
[13] Schafer, R.: 1958, 'Regression in the Service of the Ego: The Relevance of a Psychoanalytic Concept for the Personality Assessment', in G. Lindzey (ed.) *Assessment of Human Motives,* Holt, Rinehart and Winston, New York.
[14] Siegler, M.: 1977, 'Critical Illness: The Limits of Autonomy', *Hastings Center Report* **7**: 5 (October), 12-15.
[15] Siegler, M. and Goldblatt, A. D.: 1981, 'Clinical Intuition: A Procedure for Balancing the Rights of Patients and the Responsibilities of Physicians', in S. F. Spicker, J. M. Healy, and H. T. Engelhardt (eds.), *The Law-Medicine Relation: A Philosophical Exploration,* D. Reidel Publ. Co., Dordrecht, Holland, 5-31.
[16] Toulmin, S.: 1981, 'The Tyranny of Principles', *Hastings Center Report* **11:** 6 (December), 31–39.
[17] Veatch, R. M.: 1981, *A Theory of Medical Ethics*, Basic Books, New York.

STUART F. SPICKER

COERCION, CONVERSATION, AND THE CASUIST: A REPLY TO JAY KATZ

I. INTRODUCTION

The practical world of clinical medicine, like all other practical worlds, is inhabited by unacknowledged theoretical demons. A central task of philosophical reflection is to recognize and, where necessary and possible, exorcise them. Men and women in medicine typically view their work as empirically-oriented, a-philosophical, presuppositionless, practical, and free of demons and ghosts. A practical consequence of this innocent and pervasive perspective is that inhabitants of the temples of Aesculapius think philosophical reflection irrelevant and occasionally intrusive; to them it is at best a burdensome cultural knapsack – functional in a pinch, but (like most carried across campus these days) not very decorative. We cannot, however, proceed without assumptions, and philosophy is the critical examination of key assumptions. To push the simile to extremes is bad form, but one slight shove may be permitted: the knapsack, if it contains some survival gear, may at times be respected and therefore the bearing of it worthwhile. Busy with the demanding care, for example, of critically ill patients, one's philosophical assumptions are seldom worrisome to practical health professionals, who do not often think of themselves of their behavior as shaped by their unrecognized or unexamined presuppositions. Fortunately there are exceptions.

Jay Katz, Professor of Law and Psychoanalysis and philosopher of medicine, grasps the importance of principles and problems that have clear philosophical and practical implications. I believe I have located the gnawing concern which set his thought in motion on this occasion, and I shall attempt to uncover what lay tacitly hidden in his remarks. The blend of theoretical and practical concerns is clear in his presentation. It is a movement toward the center of his life's calling, and is not only a discourse on ethical or legal issues in medicine. However, the assessment which Dr. Katz directs to himself as well as to medical colleagues certainly has clear implications for moral philosophy and jurisprudence. But I shall only address myself to the former of these implications, stressing the ethical and leaving the legal implications aside. I take this

J. C. Moskop and L. Kopelman (eds.), Ethics and Critical Care Medicine, 69–77.

approach not because the legal implications of his position are unimportant, but because what is morally permissible may nevertheless be legally proscribed, and from this it simply follows that the law ought to be revised.[1] Before I turn to his "psycho-logical" progression, then, it is necessary to put the problem in perspective.

II. THE PROBLEM

It appears that sometime prior to this century, a principle emerged which Professor J. H. Van den Berg calls the "traditional medical ethical principle": "The doctor has a duty to preserve, spare, and prolong human life wherever and whenever he can ... and must fight against death under all circumstances" ([11], p. 77). This principle guided physicians until challenged during the past two decades of our century.[2] In the past two decades, however, numerous publications in the bioethics literature have challenged the principle. More recently, formidable attacks from philosophical quarters and even medical circles have led, as one might have predicted, to a backlash. Yet Dr. Katz, one should recall, pledges allegiance to the "therapeutic promise of medicine" (which I have expressed as the traditional, medical ethical principle) and this promise or principle, to which Dr. Katz adheres with unabashed fidelity, must not be broken or violated in spite of the intricate arguments of contemporary philosophers.

Can we not then view Professor Katz's presentation as a direct counterattack, by way of rejoinder, to the recent arguments of contemporary "moral enthusiasts" [10] (to borrow Toulmin's term), who have inconsistently and unreflectively defended both patient autonomy, liberty, and self-determination on the one hand *and* paternalistic overriding of a patient's freedom, autonomy, and self-determination on the other, leaving both physicians and patients bewildered, confused and, at times, angry? (Here, of course, we are referring only to competent patients who constitute a restricted set of cases of critical illness which, in Dr. Katz's words, are "illnesses that might terminate in death or lead to irreversible complications and for which diagnostic and therapeutic interventions seem to be available that might reverse or control the disease process" ([4], p. 62) and thus assist the patient to regain his/her health.) That is, physicians do not now know what is the proper *moral* justification for their decisions to override or support competent, critically ill patients who consciously refuse (1) further knowledge of their condition, (2) diagnostic tests designed to produce information directed to future treatment modalities, or (3) various forms of treatment interventions. One can perfectly well appreciate this present, unhappy state of affairs, for what should worry us most about the contemporary moral debate regarding this problem is that it appears intractable and interminable.

Consider the current moral debate and the accompanying rival moral arguments:

(1) Every *competent,* coherent patient has the right of self-determination over his own person, including his or her own body. It follows from the nature of this right that at any time such an adult patient is asked to provide an informed consent to diagnostic tests and/or treatments, the patient has a right to make his or her own uncoerced decision whether to permit the physician to perform the tests or provide the preferred treatment. Therefore, a competent patient's refusal to take tests or accept treatments is morally permissible, ought to be supported by law (as it typically is in the United States), and even imposes a duty on the attending physician to acknowledge and abide by the refusal.

(2) A *competent*, coherent patient who not only has the *capacity* for autonomous action, but who is a moral agent and therefore *autonomous,* is obliged to act reasonably if not rationally, for only reasonable or rational action constitutes a basis for the ascription of agency. In many cases it is irrational for such a patient who claims to be an agent to refuse simple and safe, though at times uncomfortable, diagnostic tests - the knowledge gained from which will likely serve, in the clinician's considered judgment, to save the patient's life and restore him to good health. Hence refusal in such cases is morally wrong, for it is in the end an irrational taking of the patient's own life by the patient himself - though it does not follow from this that a physican has the moral authority to coerce the patient by imposing knowledge, tests, or treatments on him. After all, a patient in such a condition is only slightly different from his or her normal, healthy self. Under such conditions suicide should neither be condoned nor supported in the health care context.

I believe that both of these complex arguments are logically valid or can be slightly refined as to be made so: in logician's parlance, the conclusions follow from the premises. But what of the incommensurability of the rival premises in the two arguments? Each argument contains premises that reflect normative or evaluative concepts different from the other. In the two arguments sketched here, premises that invoke *liberty* and *rights* are at odds with those that advocate *prolonging life* and *medical success*.

Can we really say that in our pluralistic society we agree on an established way of deciding between these conflicting values? The way to proceed, of course, is to work ourselves back to the premises of the two arguments. But can we, in this case, do more than invoke one premise against another, and thus simply find that we assert and counter-assert? I believe we lack unassailable criteria for resolving this particular dispute and many others like it. Thus I find no compelling reasons by means of which the libertarian can finally convince the quasi-paternalist, and *vice versa.* All this reminds one of Stephen Toulmin's defense of casuistry [10]: let the physician exercise discretion overriding a competent patient's expressed refusal to submit to tests and/or treatments, if and when the complex set of conditions so warrant. Searching for a principle grounded in libertarian freedom as an absolute side constraint, or the quasi-paternalists' overriding good intentions based on knowledge and experience, will only leave us in the position of the physician caricatured in a famous New York magazine: peering into the eyes of the complaining patient who seeks his knowledge and help, the physician remarks: "My guess is, I don't know."

III. THE PROPOSED SOLUTION

The physician who is dissatisfied with both the Scylla of libertarian arguments on behalf of patients' autonomy (though like Dr. Katz generally sympathetic that this freedom should be respected in virtually all cases) and the Charybdis of medical paternalism (which at times advocates overriding the patient's expressed refusal on the grounds that it is in the patient's overall best interests) may be struggling with an intractable dilemma. That is, adherence to an absolute principle of non-intervention *or* adherence to an absolute paternalistic, interventionist principle are both unacceptable. Moreover, the totally casuistical, case-by-case, non-principled decision procedure is also troublesome if one is seeking a solid foundation for one's decisions on more general maxims or principles. How, then, shall we make moral choices in medicine?

Dr. Katz *suggests* an interesting approach here, but one that is not philosophically complete. Let me explain.

Suppose a physician is fundamentally committed to the traditional, medical ethical principle. He or she discovers that the number of competent, critically ill patients is dramatically increasing and, furthermore, that an increasing number of such patients are, for whatever reasons, refusing to accept information pertinent to their care, diagnostic tests, or treatment. Add to this the apparently unhelpful, rival arguments generated by the philosophers and "ethicists" and the question becomes, "What are practical professionals to do in such circumstances?" One solution lies in simply working to reduce these expressed refusals significantly, but without coercing the patient or forcing oneself upon the patient (though cajoling may be permitted). After all, any form of coercion betrays the physician's failure seriously to acknowledge the patient's freedom despite agreement that in virtually all cases the patient's freedom should be respected. A physician might simply adopt a method by means of which, when skillfully applied, such patients rescind or, better, never even voice their refusal. In short, Dr. Katz appeals to a form of *conversation* which breaks the paralyzing silence that sometimes lies behind such refusals. Presumably, when one comes to appreciate the elements of Dr. Katz's "clinical theory" ([4], p. 43) of human interaction, one will have disclosed the structures of our "psychological nature" ([4], p. 53) and bridged the chasm that currently separates the "appeal to moral principles" on the one shore and the "casuistical approach" on the other.

Though I am convinced that Dr. Katz's clinical theory is not adequately explicated in his remarks - since he does not tell us how the theory is based on "psychoanalytic propositions" and the assumptions of human bio-psychology - in fairness he does mention what I take to be some fundamental considerations which are required of such a theory. The starting point, it appears, is the distinction he draws between autonomy as a *moral principle* and autonomy as a *psychological capacity* ([4], pp. 53–54), the latter being the actual set of capacities (1) to choose among alternative courses of action; (2) to think, deliberate, or reflect; and (3) to act as an agent in human, concrete contexts – in this case the health care setting.

What we need, though, is (1) a far greater explication of this clinical theory, (2) a clearly articulated description of the psychology of human nature and (3) a useful account of what constitutes a "radically different kind of dialogue between physicians and patients" ([4], p. 66). Perhaps these will be forthcoming in later writings

At this point, I wish to return to Dr. Katz's commitment to the *mutuality of conversation* between patients and physicians which, in his view, is too often exploited by physicians who should know and do better. He calls for mutual respect for each other's vulnerabilities and incapacities - usually the patient's infirm *physical* condition and the physician's *psychological* condition, a thesis, by the way, that is reminiscent of the now classical Szasz/Hollender essay of 1956 [9].

I find this suggestion intriguing and one that clearly calls for another *kind* of active relation between patients and physicians.[3] But if very special skills are needed to enable respectful conversation to take place, then I am totally puzzled when Dr. Katz says, as he does, that "I am not suggesting, however, that the deliberations between physicians and patients be converted into an exploration of the psychological roots of patients' and physicians' motivations and expectations. This is neither warranted nor possible" ([4], p. 59). It seems to me that it is precisely such detailed psychological explorations that offer any hope of realizing authentic and respectful conversations, which will not require the physician to choose among either libertarian or paternalistic principles on the one hand or case-by-case casuistical discretion for exceptional cases on the other (that is, cases where it is morally permissible and perhaps even obligatory to override a patient's refusal). Futhermore, Dr. Katz's position seems to require that physicians make a very serious study of their own irrational and unconscious preferences and values. How else can physicians determine the patient's and their own "underlying psychological incapacities?" ([4], p. 61).

In sum, Dr. Katz's impatience with bioethicists and physicians compels him to seek the practical ends noted above: to make the problem of stand-off between physicians and patients a "rare event" ([4], p. 61) and to reduce the frequency of patient refusals to as near zero as possible so they are "a rare exception" ([4], p. 65). We are reminded that should this be the case, then (in lawyers' parlance) these issues simply become moot. The professor of psychiatric medicine emerges as the professor of jurisprudence!

Finally, I hope that Dr. Katz will admit that even physicians with the best intentions and psychotherapeutic skills will encounter some patients who continue to refuse to engage in further conversation and insist on their prior convictions. Moreover, some patients may, precisely because they have participated in respectful conversation, first decide to refuse diagnostic tests or treatment. This is an important possibility and for some even a paradox.

IV. AFTERWORD

As I was preparing this comment on Dr. Katz's essay, a colleague, Professor Larry Ulrich, brought an article to my attention that appeared in *The New England Journal of Medicine* on September 2, 1982, written by Dr. Mark Perl and Professor Earl E. Shelp [5]. The article warns of the dangers that exist when psychiatric consultations are requested in such a way as to mask the moral issues which are really at the center of the attending physician's request. The authors' argument seems plausible enough and they offer three cases to illustrate their concern. In spite of this, their general thesis is unsound, for they expect us to believe that a sharp line can and should be drawn between a psychiatric and an ethical problem and that the tendency of psychiatrists to generalize their expertise, such that they be viewed as competent to undertake ethical analysis, is at present a case of *hubris* in medicine.[4]

Having attended numerous grand rounds which included psychiatric evaluations, I think the thesis of Perl and Shelp is overstated. In point of fact it is much easier to argue, and to show, that psychiatric consultation is an *intrinsically* value-laden enterprise for both patient and psychiatrist. Though I would not wish to deny that conflicts between competing moral values generate ethical dilemmas (indeed, I have argued that they are typically intractable and reducible to rival premisses for which no final arguments appear decisive), I *do* believe that psychiatrists are trained to explore the moral and value-laden aspects of a patient's problems - including those sticky ones where patients refuse further diagnostic tests and/or treatments. In fact, Dr. Katz's entire essay rests on his willingness, *qua* physician, to participate in "respectful conversation" so that the moral and value-laden aspects of his patients' concerns are confronted directly. He has, after all, taken pains to underscore the distinction between autonomy both as a *capacity* and as a *principle* of freedom.

Perl and Shelp in fact agree with Dr. Katz's point. They remark, in concluding their essay, that "... the psychiatrists' mission has traditionally been one of helping patients arrive at decisions *autonomously,*[5] of providing the opportunity to discuss and explore complex issues and feelings in a nonjudgmental setting" ([5], p. 620). How, one might ask, could any psychiatrist undertake such a task without some degree of skill in moral analysis? The point is that medicine, especially psychiatric medicine, is an intrinsically ethical enterprise (i.e., as even Perl and Shelp maintain, when they point out that psychiatrists often assist patients in arriving at decisions "autonomously") in which philosophical argument, when it is useful for clarification of certain issues and concepts, is frequently an ingredient in psychiatric conversation. In the end, however, what is usually more important is the humaneness and compassion of the interaction and the avoidance of inhuman coercion on the part of the physician – not the degree to which a humanistic, intellectual ideal emerges from respectful conversation and is, alas, thwarted or realized in the end.

University of Connecticut
School of Medicine
Farmington, Connecticut

NOTES

[1] Before proceeding further one tactical point should be made: Professor Katz has often referred to three authors - Eric Casell [3], Mark Siegler [6], and Stephen Toulmin [10], the former two being physicians and the last a philosopher. As none of these men have the opportunity in this volume to respond to challenges posed by Dr. Katz, I shall not presume to speak for them, except to say that in general Dr. Katz agrees with their main theses. This to some degree explains why it is proper to view Dr. Katz's concerns as partially hidden – his calling card, like Descartes's, being *larvatus prodeo* – "I come forward in a mask." So I shall not here attend to Dr. Katz's critique of these contributors to the bioethics literature. (If space permitted, I would try to show that Dr. Siegler's arguments in the two essays cited by Dr. Katz [6, 7] reveal that his actions with respect to the patient, Mr. D., were not consistent with his theory of clinical intuition. Nevertheless, it is more important to view Dr. Siegler's inconsistency as a symptom of the difficulties which moved Dr. Katz to exercise us as he has. A careful reading of [7] will reveal that Dr. Siegler's case of the patient, Mr. D., which took place some four years before (see [6]), on the criteria noted in [7], would have led to a decision *not* to acknowledge and accept but to *override* Mr. D's refusal of diagnostic tests and treatment.) See especially [7], pp. 15, 19 and 23.

[2] Professor Amundsen soundly argues, by way of a careful review of ancient medical-historical texts, that it may be more contemporary myth than accurate reconstruction of Hellenic medicine, that in classical medicine the physician had an unqualified duty to prolong life ([1], p. 28).

[3] Dr. Katz's view is shared by other contributors to the bioethics literature. In 'Allocating Autonomy: Can Patients and Practitioners Share?' Sally A. Gadow writes: "The role of the professional in self-care ... is not merely to *protect* patients' freedom of self-determination against institutional encroachments, inadequate information, or professional prejudices. It is far more than that. It is *active assistance* to individuals in their exercise of self-determination. It entails participating with patients in discussion that has as its mutually agreed upon purpose the discernment of the patient's view and values and the examination of health care options in the light of those" ([2]. p. 101).

[4] The matter of determining whether physicians and humanists are at times guilty of the informal fallacy of "generalizing their expertise" without warrant, was briefly debated by Robert M. Veatch and the author in November, 1978. At that time the medical context was the moral issues generated by the work of Dr. John Lorber in his policy of selecting certain newborn infants for non(surgical) treatment of myelomeningocele (see [8], pp. 212-213, note 12).

[5] My emphasis.

BIBLIOGRAPHY

[1] Amundsen, D. W.: 1978, 'The Physician's Obligation to Prolong Life: a Medical Duty without Classical Roots', *Hastings Center Report* **8**: 4 (August), 23–30.

[2] Bell, N. K. (ed.): 1982, *Who Decides?: Conflicts of Rights in Health Care*, Humana Press, Clifton, New Jersey, especially Chapters 7–10.

[3] Cassell, E. J.: 1977, 'The Function of Medicine', *Hastings Center Report* **7**: 5 (October), 16–19.

[4] Katz, J.: 1985, 'Can Principles Survive in Situations of Critical Care', in this volume, pp. 41-67.

[5] Perl, M. and Shelp, E. E.: 1982, 'Psychiatric Consultation Masking Moral Dilemmas in Medicine', *New England Journal of Medicine* **307** (10) 618–621.

[6] Siegler, M.: 1977, 'Critical Illness: The Limits of Autonomy', *Hastings Center Report* **7**: 5 (October), 12–15.

[7] Siegler, M. and Goldblatt, A. D.: 1981, 'Clinical Intuition: A Procedure for Balancing the Rights of Patients and the Responsibilities of Physicians', in S. F. Spicker, J. M. Healey, and H. T. Engelhardt, Jr. (eds.), *The Law-Medicine Relation: A Philosophical Exploration*, D. Reidel Publ. Co., Dordrecht, Holland, pp. 5–31.

[8] Spicker, S. F. and Raye, J. R.: 1981, 'The Bearing of Prognosis on the Ethics of Medicine: Congenital Anomalies, The Social Context and the Law', in S. F. Spicker, J. M. Healey, and H. T. Engelhardt (eds.), *The Law-Medicine Relation: A Philosophical Exploration*, D. Reidel, Publ. Co., Dordrecht, Holland, pp. 189–216.

[9] Szasz, T. and Hollender, M.: 1956, 'A Contribution to the Philosophy of Medicine: The Basic Models of the Doctor-Patient Relationship', *AMA Archives of Internal Medicine* **97**, 585–592.

[10] Toulmin, S.: 1981, 'The Tyranny of Principles', *Hastings Center Report* **11**: 6 (December), 31–39.

[11] Van den Berg, J. H.: 1969, *Medische Macht en Medische Ethiek,* G. F. Callenbach, N.V. Nijkerk, Holland.

LORETTA KOPELMAN

JUSTICE AND THE HIPPOCRATIC TRADITION OF ACTING FOR THE GOOD OF THE SICK

In this paper I will examine whether the construction of a just system of health care is impeded or assisted by a Hippocratic tradition, the duty of beneficence. It will be argued that it is an important part of a just system of health care that seeks empathy, and equity as well as impartiality. In the *first* section the charges that the duty of beneficence promotes unwarranted partiality and paternalism are examined. It is found that this tradition understood as a prima facie or imperfect duty of beneficence, neither entails nor fosters unwarranted paternalism and that this moral duty may be important to a just system. Partiality, for example, when it results in empathetic care, may be defensible in a just system. In the *second* section the reason why this kind of duty may be important in a just system is explored. Providing care for individuals, rather than thinking of what faceless members of some group need, encourages personal regard and discretion.

In the *third* section it is argued that especially when resources are scarce and in demand, the particular society must take some responsibility for clarifying the proper scope and limitations of partiality or paternalism. Professional and personal commitments may suggest competing views about the allocation of scarce resources (for example, each wanting to care for his or her own), and without a defensible general policy the choice between the individuals recommended for care may seem arbitrary. The strain on such a system is greatest in tragic situations when great harm comes to some who cannot be helped. Systems of justice are, as Hume pointed out, constructed for conditions of moderate scarcity in the allocation of goods and services [22]. But in allocation of scarce but critical care in medicine, conditions of moderate scarcity are not met. Situations, then, are inescapably tragic if because of the scarcity of a life or death resource, some who could be helped cannot be helped [29]. While seeking social understanding about limitations and expectations seems important and reasonable, there may be a social reluctance to recognize that moral dilemmas exist and that tragic choices must be made. It is argued that the Hippocratic rule to do the best you can for your patient, or to act for the good of the sick, is a reasonable prima facie

J. C. Moskop and L. Kopelman (eds.), Ethics and Critical Care Medicine, 79–103.

moral rule, but when resources are limited and it is transposed into "everyone ought to get the best," then it becomes an incoherent and thus unjust social policy. I will begin by clarifying what is meant by the Hippocratic duty of beneficence and what problems it seems to create concerning fairness of delivery of health care.

There is an ancient precept in medicine that one has a duty to try to act for the good of the sick either by preventing harm to them or by benefiting them. This duty of beneficence will here be called a Hippocratic tradition because of its importance and similarity to a portion of the Oath of Hippocrates:[1]

> I will apply dietetic measures for the benefit of the sick according to my ability and judgment; I will keep them from harm and injustice ([21], p. 121).

This passage does not say who should be helped if all cannot be helped, or how far one should go in trying to benefit the sick or keep them from harm or injustice. Although some view acting for the good of the patient as "the most ancient and universally acknowledged principle of medical ethics" ([32], p. 117), others charge that this moral rule encourages unwarranted partiality (failure to consider the needs and interests of the many by focusing too exclusively on the needs or interests of the few) and unjustified paternalism (failure to respect the liberty of others) on the part of those adhering to it.

Consider how problems of partiality and fairness seem to arise from two important readings of this moral rule. It may be understood as advocacy for the interests of either an *individual*, especially a patient under one's care, or many or all patients *collectively*. But since what is good for one may conflict with the interests of others, these interpretations can recommend different actions. When, for example, needs are great and resources limited, and all cannot be helped who could benefit from aid, then partiality to one's patients can make a difference to others. If health professionals understand this moral rule as a vindication of special commitments to their own patients or as suggesting that their own patients' needs should have a greater claim to their attention than strangers needing quite similar services, then they may be partial. Since such professionals often help select who gets care when not all can be served, the needs of the stranger may be unfairly weighed if professionals, in their role as advocates, seek to do their best for their own patients, including capturing limited funds or space for them before it is given to others. Thus, one problem is: Do we risk being unfair by allowing partiality, or do we risk sacrificing trust and empathy in the therapeutic relationship if we do not?

Another charge is that this duty of beneficence entails or fosters unwarranted paternalism. Even if we suppose that physicians since Hippocratic times have generally been unenthusiastic about sharing important medical decisions or inviting patients to participate in them [23, 27, 32, 33, 46, 47], is this a moral rule to which they can legitimately appeal to vindicate what would otherwise be unwarranted paternalism?
understand this Hippocratic tradition in relation to some general requirements of justice. This Hippocratic moral rule has found a place in societies with different solutions to allocation problems. Cultural goals, values and priorities can lead to justifiable differences in allocation. For example, a nation at war may give its army first priority when distributing a scarce medical resource as was done with penicillin in World War II. Reasonable particular solutions to problems of justice, then, contain some elements *contingent* upon times and circumstances. It would be an endless task to examine the fairness of this Hippocratic moral rule in relation to each social arrangement in order to make some general remarks about its fairness. Examining it in relation to *formal,* or necessary features of justice, simplifies our task and two formal requirements of justice will be used.[2]

It will be assumed that *any just system* fulfills two necessary but not sufficient conditions:[2] *First,* a just system requires the equal treatment of equals, unequal treatment of unequals but in proportion to their relative differences. This is a formal requirement for consistency in treating alike those similarly situated [2, 28]. *Second,* a just system is practical and encourages reasonable and informed persons of good will to be just and to act justly [2]. That is, informed people of good will can live with and use the system, or enough of them can to keep the system viable. For example, a system punishing those disagreeing with state policy with invasive "therapy" and a diagnosis of 'psychosis' would be a system informed people of good will could not use or accept.

Few problems are likely to arise for any system where everyone agrees or where there is no scarcity. The strain will be greatest when the decisions are controversial or urgent, and goods and services are scarce. In critical care medicine, decisions can be like this. They are often attended by the knowledge that help cannot be given to all in need and that the consequences of doing without are a tragic loss of life, health or opportunities. In such circumstances, overriding a patient's foolish choice seems appealing as does showing partiality for those to whom one has professional and personal commitments. Since the temptation to be paternalistic or show favoritism may be great when resources are scarce or care is critical, our discussion will focus on this.

To introduce the importance of contingent as well as formal requirements of justice, consider two different solutions to allocating critical but scarce medical resources. The first one seeks *impartiality*. A detailed and agreed upon computerized formula is used to select patients. This standard might be based upon maximizing the likely medical benefit to all in need given their respective prognosis; or some random technique to select between those who are equally in need; or some consideration of cost to likely health benefit; or merit; or social utility. The essential feature of the system is that it leaves little or no room for favoritism on the part of those distributing the scarce goods or services. Such a system might meet our first necessary condition of a just system, treating those similarly situated similarly. But the test of a system should also include how well it can be used by those for whom it is intended. This is the second necessary condition of a just system. Suppose there is genuine hope for recovery for a long-cared-for friend or patient who is asked to leave an intensive care unit because the computer print-out indicates a stranger ranks somewhat higher. Would such a system relieve professionals of the burden of having to decide in particular cases or would it impose the greatest burden on those who, in caring for their patients, care about them as well? What would be the loss if this sort of impartiality discourages personal attention to and concern for patients as individuals by making empathy for patients on the part of professionals painful, or by requiring that, in fairness, people must be told when allocation choices unfavorable to them are made? Whether or not this kind of system would inflict a great hardship on those who have to use the system, it probably could not succeed without wide social and legislative support. There is now a significant legal risk to staff and hospitals if they withdraw therapy at all or in favor of another therapy when patients or families disagree. If laws and professional codes support staff and institutions giving full consideration to "their own" patients, then this is likely to encourage partiality and discourage balancing the relative prognosis, needs or costliness of the many others who are sick.

This illustrates the importance of a social consensus in determining solutions, and also suggests another view about how to allocate scarce medical goods and services. We may adopt a policy of *partiality* where rights and duties are carefully prescribed in order to favor as full a commitment to the good of patients under care as possible; the needs of others remain outside one's jurisdiction until professional or institutional responsibilities have been formally assumed. The needs of the acutely ill stranger, then, with a good prognosis seeking admission, would be ignored in favor of the dying person in the unit when only one can be helped. This second solution may not satisfy the condition that it treat similarly situated persons similarly unless one considers the only relevant feature to be that of being someone's patient at the right time, or first in line for the service at an institution.

The first solution takes the mark of a just system to be giving each his due by imposing on all rigid impartiality in distributing goods and services; the second takes it to be giving each his due by performing exactly certain tasks or fulfilling contracts, irrespective of the needs of the many. In contrast to both, however, some systems of justice allow discernment in applying rules, openness to the plight of others and encouragement of responsible choice. Thus, the relation between justice and impartiality, or how flexible we want our system to be, emerges as an important consideration. If in different times and places alternative answers are reasonable, then it is a contingent issue as well. These examples suggest that in framing actual systems, the nonformal requirements of justice, such as what is thought scarce or how much partiality is tolerated, are important. They also show the role of social consensus. This does not mean that everyone will agree, or that social approval makes something correct, or even that a consensus is always necessary. The character of the system, however, will be affected by contingencies such as specific goals, values and priorities. What is considered useful, thought scarce, perceived to be rational options for action, regarded as a duty, likely to be funded, or supported by law, will influence choice. Nonetheless, some socially approved arrangements are unfair as, for example, an allocation system excluding a powerless minority. Our general problem, then, is: should a fair system honor this Hippocratic duty of beneficence?

I. CRITICISMS OF THIS HIPPOCRATIC TRADITION

Moral duties are typically understood as having a *ceteris paribus* clause or as holding, "all other things being equal." It is here argued that the precept to act for the good of the sick should be understood in this way. Thus, one could spell out the rule: *All other things being equal one ought to try to act for what one believes to be the good of the sick either by preventing harm to them or by benefiting them either collectively or individually.*[3] As a practical guide for action, then, the rule should instruct us to act for what we believe to be someone's good, all things being equal. In this section we will examine whether this tradition thwarts basic requirements of justice because of unjustifiable paternalism or because of unwarranted partiality as some have charged.

A. *The Charge of Paternalism*

The moral rule under discussion incorporates two kinds of notions, keeping the sick from harm and benefiting them. Either may be an unwarranted intrusion if one is competent and does not want to be benefited or protected.

Some draw on writings from the Hippocratic corpus to show, historically, that paternalism was favored by physicians of that time [23, 47]. One could maintain, moreover, that the 1980 code of the American Medical Association in breaking with past traditions by including notions readily identified as opposed to paternalism, such as dealing honestly with patients and respect for the rights of patients, indirectly admits that unwarranted forms of paternalism had been favored [1]. Others simply define this tradition in a way that is paternalistic. Veatch attacks a medical paternalism based on the principle that "the physician should do what he thinks will benefit his patient (even if the patient does not happen to concur)" ([46], p. 7). Although Veatch calls this position 'Hippocratic', it will be argued that it is not entailed by the moral rule given above or presented in the Hippocratic Oath.

Paternalism is sometimes justifiable when persons are disabled or incapable of acting on their own behalf or when their reasoning is impaired by ignorance for which one is not responsible. Paternalism is not only permissible but sometimes a duty when people: cannot make decisions for themselves; or are in the grips of incapacitating mental illness; or show involuntary, self-destructive behavior; or make choices so inappropriate to their own life-long goals that we doubt their autonomy [51]. In the presence of non-autonomous, self-destructive behavior or when means are used that are irrational, unreasonable and uncharacteristic, then paternalism may be justified.[4] Determining that the act of the moment is intended, then, is not a foolproof test of autonomy.

The moral duty to do what one believes is good for the sick would be unjustly paternalistic if it stated that *all* one has to do is decide what one believes is in the patient's best interest and then do it. Then it would not matter what patients, families, or peers thought, or what the laws or the latest journals state. Believing one is doing the best one can, however, at least shows sincerity of purpose. This is an important consideration, for it may never be correct to go against one's conscience. A competent adult may have a right to starve herself but no right to require her appalled physician to help her. One's conscience, however, may be informed or not, prejudiced or not, empathetic or not, or partial or not. Thus doing what one thinks is best is not sufficient to justify action. If this were what is stated by this duty, or by the Hippocratic Oath, it would be indefensible. This is not what is said. Medical traditions, including the Hippocratic Oath and corpus, hold that not only is it necessary to do what one believes is best for patients, but that one must keep up one's skills, study, respect confidentality and consult appropriately with peers. While we may disagree with some Hippocratic instructions even finding some of *them* paternalistic, their existence shows that one's notion of what is best for the patient does not and was not meant to settle what ought to be done. This moral rule, then, was understood even in ancient times as *either* a test of conscience of the virtuous practitioner, *or* as a rule of action having a *ceteris paribus* clause or as a *prima facie* duty.

Thus, the rule itself should not be taken as an absolute one. Lacking, and only recently remedied in such codes [1], however, is *another* norm stating that patients ought to be consulted about their care. The absence of one important rule, however, does not necessarily undermine the importance of another. To say that it does would be like criticizing this norm for not also clarifying its relation to community health, research or prudent relations with colleagues.

Some may wish to argue, however, that even if there is no entailment between this Hippocratic tradition and unwarranted paternalism, it creates the wrong sort of attitude. It encourages physicians or staff to exert pressure based on what they happen to think is best. For example, when informed that the prognosis is grave without treatment, a legally competent patient may refuse the recommended treatment for reasons that seem obscure or irrational. Staff then persist in their efforts to convince the patient he must have the needed therapy for his own good. Why is it assumed that this tendency of staff, if not carried to the point of manipulation or harrassment, is paternalistic? Perhaps it is commendable for them to be concerned. Critics, however, seem to find it presumptuous for staff to assume they know better than the patient what is best for him. Staff, critics say, act disrespectfully in trying to convince the patient of another course because of their own uneasiness about the decision, failing to see patients as beings on an equal footing who are equally capable of assessing what is in their own best interests.

But this argument is less convincing once the distinction is made between acting *intentionally* (one knows what one does and is able to do otherwise) and acting *autonomously* (one is self-legislating, in control and responsible for one's intentional actions as a free informed person). Depressed people may choose not to eat. It may be correct to say they act intentionally. But when we say of that person "He is not himself" or "This is not like him" we mean that his actions, though intentional, are not autonomous.

Some decisions which seem to us tragic are autonomous, such as the refusal by an eighteen-year-old Jehovah's Witness of a life-saving blood transfusion. Excepting those that cause harm to others, choices by competent informed adults generally must be respected or assumed to be autonomous. But when we believe the choice is determined by tides of fear, anxiety, depression, pain or misconception, then we suspect that the decision, while arguably intentional, is *not* autonomous. It seems more like an impaired choice than one that is free and informed. This suspicion is not rare when dealing with people who are sick, suffering, and faced with crucial choices, but who act in an unfathomable manner.

Even if we ultimately should respect the free choices of all competent adults when they do no harm to others, there is often a residual worry where the choice seems unfortunate. If we are concerned about or respect others, we not only worry whether *they* are really competent, acting voluntarily or aware of what is at stake but also whether *we* could have done a better job of helping them to see what is important. Such practical concerns are not necessarily acts of unjustified paternalism or unwarranted interference. The *right* of competent adults to do foolish or tragic things need *not* be the issue when our worry is whether they really understand what is at stake. Discussing and trying to understand things from each other's viewpoint is, Rawls argues, showing respect [36]. One possibility we should recognize is that they may convince us we were wrong [23].

A distinction is commonly made between *strong* and *weak* paternalism. Overriding the competent choice of competent persons "for their own good" or "to prevent their harm" is the act of a strong paternalist. In contrast, a weak paternalist intervenes to determine whether the person's choice is impaired or not. Fear, ignorance, medication, illness, or undue pressure may impair the judgment of an otherwise competent person. Weak, unlike strong, paternalism could be a sign of respect and concern if it promotes a competent choice. That is, it may be more respectful than either overriding the unimpaired choice of a competent patient or abandoning the person to his decision without inquiring if it is a competent, voluntary and informed choice.

The legal consent requirement, then, is not the entire or often even the central issue. Staff may know perfectly well they have fulfilled their *legal* obligations to people and yet, because of the personal or professional commitments to try to do the best they can, believe they should do more. They may know that they have given all information a reasonable person would consider material for the decision, that the person is legally competent, and that undue pressure is not being applied. The rules, minimal standards, or legal requirements may have been met long ago, but conscientious staff may be frustrated, believing that the choice (albeit intentional and by a competent person) is impaired by irrational views, fear or ignorance. Their worries or attempts to explore ways to see if patients *really* understand, act voluntarily or are competent to consent, need not be a sign of unjustified paternalism, but of care about the individual.

Disrespect to others can take many forms such as: unwarranted paternalism, indifference or abandonment of others to an impaired choice, manipulation of information, strategies endlessly to assess whether the disliked choice is competent, or other means of harassment.

Of course, people may improperly seize upon this Hippocratic tradition (or most anything) to rationalize imposing their views on others. *Saying* one is acting in someone's best interest, when you really want to do something else, is different from *trying* to act in someone's best interest. And both saying and trying are different from trying to do so when sufficiently *informed* about the relevant data. Some people inappropriately assume they know what is best for others. For example, it is arrogant and unwarranted to claim, as some do, that they know without consulting the patient or family that some highly invasive, disorienting and painful procedure that will briefly extend a diminishing life is in the patient's best interest.

It seems ironic that the health care team, sometimes rightly criticized for cavalier paternalism and encouragement of dependence, is also typically necessary for most patients or families to exercise autonomy in important medical decisions. Generally, if the patient's condition is acute and serious, then unless the professionals are prepared to be very helpful when informing patients or families, they may have little genuine role in the decisions. Being adequately informed may be simply understood as being served enough information to satisfy some legal requirement. But it is hard to assimilate bad news and often people do not "hear" the information when they are initially "informed." Those determined to do their best to inform others of newly acquired, serious, chronic or life-threatening illness must be prepared to spend a lot of time with them. Their legal obligation could be discharged by presenting sufficient information for a reasonable person to understand and make a decision [6, 24]. And, a signed consent form which includes that information is evidence it has been presented. Those who go through complex or upsetting material repeatedly and on different occasions until the patient or family grasps it, however, can take on a time-consuming burden, not for their own legal protection, or to make their own duties easier, but to make autonomy a possibility for many patients or families. If they do so because they try to act for the good of the sick, then they are using the moral rule we have called a Hippocratic tradition.

B. *The Charge of Partiality*

The second charge is that this Hippocratic tradition of pledging beneficence unjustly encourages people to ignore or neglect the needs and interests of the many by focusing on the needs and interests of the few, namely one's patients. Consider the following argument which appears to generate a contradiction between this Hippocratic moral duty understood as a duty to favor one's own individual patients, and a necessary condition of a just system.

(1) A just system requires the equal treatment of equals, unequal treatment of unequals but in proportion to their relative differences.

(2) In adopting a Hippocratic rule of action to do the best one can for one's own patients, one places the needs and interests of some above others irrespective of important relative similarities or differences.

Thus, (3) In adopting a Hippocratic rule of action to do the best one can for one's own patients one unjustly places the needs and interests of some above others irrespective of important relative similarities and differences.

The second premise is controversial, however, if we recognise some special regard as important and reasonable in a just system. For example, we regard being-one's-parent as an important consideration, thinking it proper to take more interest in the needs of our own parent than a stranger's. The issue, then, is whether being-one's-patient, like being-one's-parent ought to make a difference in deciding who should get things from us that cannot be given to all. The initial problem, then, is that not all can be helped.

This problem is a frequently discussed kind of moral dilemma: Who should be helped when only some can be helped? [28, 29, 25] Those who help one person, and by that act cannot help another or others similarly situated, select what ought to be done from among two or more justifiable acts. Where resources are scarce, doing one's best cannot mean doing everything for everyone with similar needs or interests, since that is impossible. In general terms we express this as *ought implies can*. As circumstances change, what we can and ought to do may change as well. And sometimes we *do* think it proper to help strangers and let friends or patients be inconvenienced. For example, breaking a promised appointment to avoid a great evil is generally regarded to be appropriate. This may be because of our compassion or because we hope others would reciprocate and do that for us. Suppose a physician rushes past a desperately needy physicianless patient to meet a comparatively well private patient, claiming that he acts for the good of his own patient whom he has promised to meet. This reasoning fails if the rules to keep promises or act for the good of the sick, like most or all other moral duties, are not absolute but *prima facie*; if so, then adjustments or ranking of such duties must be made. If given an opportunity to do a great service, it seems appropriate for the physician to break this promise to the private patient, and favor the physicianless patient. Furthermore, it cannot be an absolute duty to do one's best for each patient, if one has more than one patient or competing duties to oneself or others.

Several important reasons can be given to show that this Hippocratic tradition to act for the good of the sick is different from a system aimed at an entirely impartial distribution of goods and services. *First,* one can provide care, show compassion or be responsible for some or a few people, but this is not the same as being responsible for allocation of resources to all sick people, all those who might need a certain facility, or all who have similar conditions when resources are scarce. To criticize as partial those who try to help some but cannot help all is to blame them for what is unavoidable. When those providing care cannot hold out help to everyone similarly situated because needs exceed resources, such criticisms assign unreasonable responsibility and deny the initial problem of scarcity. *Second,* hope and despair can be self-fulfilling prophesies such that if hope is realistic it can help determine a good result. Thus, the health professional - more like a coach than an impartial judge - may encourage optimism in doing the best for someone. *Third,* the policy of close impartial comparisons requires something that is at least extremely difficult and perhaps impossible. It is difficult since there is often genuine uncertainty about an individual's prognosis even where there may be clear group trends. Prediction relates to groups and likelihood of individual response. Individual constitution, determination, and hopes and fears are among the things affecting outcome, but they are elusive variables for prediction. If close judgment between people requires fair and reliable comparisons, it may for all practical purposes be impossible. The important features of the patient's attitudes, environment and constitution may not be known or even knowable for such a purpose. Indeed such comparisons may be theoretically impossible. To put a complex matter briefly, if beliefs, hopes, despair are (1) only expressible as intentional claims, (2) not reducible to or substitutable for extensional or scientific claims [8], and (3) important in determining prognosis; then there is no theoretical way to include in the extensional calculus all significant variables. But this is what is being asked of a purely impartial system that would include significant variables such as those hopes, beliefs, attitudes and fears relating to prognosis. *Fourth,* people to whom care has been provided are no longer strangers. Help is given to individuals who then presumably ought to concern us as individuals. Decisions about what is strictly owed to them, however, suggest regard for them not as *persons* but as members of a *class* of people. for example, those who are in need of the unit's space.

Thus, the Hippocratic moral rule seems compatible with the formal requirements of justice enumerated, but different from (and it will be argued generally preferable to) a *purely* impartial system. If we seek justice allowing personal attention and discretion in its application, then it too may be different from a purely impartial system. Although partiality when the result of empathetic attention to people may be defensible, scarcity creates strains. In tragic situations such as in critical care medicine when help cannot be provided to all who need it, the limitations of partiality need to be clarified. But in this respect, the problem of who to help when not all can be helped is like some other moral dilemmas. We shall examine how the requirements of compassion and justice, generally supportive frameworks, may sometimes suggest different courses of action.[5] These two ways of determining what ought to be done are reminiscent of the singular and collective interpretations of the Hippocratic moral rule. Those concerned for the good of a particular patient employ the *individual* interpretation of the Hippocratic moral rule; those concerned with all or many patients employ the *collective* one.

II. IMPARTIALITY AND COMPASSION

In the last section it was argued that attention to and compassion for people are important to understand the Hippocratic tradition to help those who suffer. Impartiality and compassion, however, sometimes seem different: not mutually exclusive, but like two games that are hard to play simultaneously [25]. In a just system, compassion and impartiality each seem important. Their relationship has been explored by philosophers, including Aristotle, Hume, and Weil, who argue that there is a necessary relation between promoting compassion and the creation of a just system [2, 22, 50]. Weil holds that from compassion we learn of equality and the meaning of rights [50].

In this section we will briefly examine the relationship of compassion, impartiality and justice. *Compassion, empathy,* and *friendship* are here understood to require attention to an individual. They are distinguished from *impersonal sorts of charity, altruism, benevolence or social justice* that can be practiced without worry for someone in particular 22, 34].

Empathy or compassion, herein used interchangeably, require attention to an individual and imagining his or her situation [2, 22]. It is easier to ignore the plight of the unseen than seen, the unheard than heard, the untouched than touched [30]. The role of imagination in all empathy, the possibility of its even supplying the person attended to, and the difference between personal and impersonal charity are brought out by Evgeny Evtushenko:

> Even the lowest of the low feels pain. His own, of course. The aching done for your fellow man is called empathy - and it is the highest form there is. A rare gift. But obligatory for the writer of stature - the greater the scope of his empathy, the greater his own. Pity for the frozen bird, for the faded flower can ignore the tragedy of a people, if not all mankind. Or cry in loud and abstract tones for the fate of a planet without bothering to pick the sufferer up ([14], p. 7).

What is the relation of empathy to justice and how may they suggest different actions? Two of the best known philosophical contrasts between the morality of friendship, compassion, charity or love on the one hand, and justice on the other are those of Aristotle and Hume [2, 22]. Clarifying these two different ways of deciding what ought to be done will help us understand the location of a Hippocratic tradition of service to the sick within a just and equitable health care system.

Compassion or friendship typically guides us in deciding what we ought to do for friends, those we love or even those who have our attention because they ask for our help. In our concern for them as individuals, we do not normally consider schemes of fairness, consistency, impartial application or rules, allocation of resources or other issues of justice. Rather, we care about their needs or desires *as individuals*. Through empathy the joys and sadness of others become our own. The demands of this kind of sympathy can be so emotionally draining that professionals could not provide it for long to many; students in medical professions, therefore, use 'empathy' and 'sympathy' as technical terms and are taught to distinguish between the requirements of empathy (a caring but professional response) and sympathy (caring as a friend without professional bounds).

Though we benefit from and need compassion, actions that exemplify compassion like kindness, gratitude, and affection are characterized by a concern for another or others, rather than self-interest. Although

Aristotle wrote that friendship "seem[s] to lie in loving rather than being loved" ([2], 1159A26-27), in general, he regarded friendship as requiring equality and reciprocity. If we reserve friendship for relationships which depend on equality and reciprocity, as Aristotle did, then compassion or love is the broader category, and the parable of the Good Samaritan is an instance of love and not friendship. If, like the Society of Friends, however, we mean by 'friendship' the wishing of another or others well, then reciprocity is not an essential feature. One can show this kind of friendship, or compassion, without expecting a similar return. For example, to get empathy patients do not have to empathize with the sufferings of the staff (though it usually helps). All compassion or friendship requires feeling for and with another and is marked by attention to the person rather than just to the class to which the individual belongs.

Hume wrote that a deed based upon compassion and friendship is typically direct and uncomplicated and "chiefly keeps in view the simple object, moving the affections, and comprehends not any scheme or system, nor the consequences resulting from the concurrence, imitation or example of others" ([22], p. 16). We help our friends and comfort our children because we care about them; we do not feel committed because of this to treat everyone similarly situated the same, nor do we reflect on the resources of the community, state or world before we act. He argued that such deeds of kindness have great personal and social benefit in promoting good will and instructing us about creating a virtuous life [22]. But he warned that while important to justice, deeds based upon friendship and compassion are different. The following example shows this.

It is not unusual for staff to become fond of someone who has a long hospitalization. One child, born quadriplegic, later found to be severely retarded, spent her first two years of life in the intensive or special care facilities in a hospital. Charges were made that some staff had "lost perspective" and "invested their own fantasies" in caring for the child. But from Aristotle's and Hume's views of the nature and importance of compassion some of this was understandable. For on their analyses, in feeling compassion, the joys and sorrows of individuals become our own. The issue, then, is not the fantasies per se, but whether duties to others have been set aside unjustly.

In wanting to help a sick individual, staff may ignore issues of billing, resource allocation or what can justly be provided to all those who are equally sick. Rules of what is strictly owed to everyone similarly may seem inappropriate when compared to a natural and immediate affection for those we care for or about. In contrast, those attending to bills, cost – effective use of beds, or providing care first to those in the district use a different framework.

Impartiality. In a just system impartial treatment is sought and special treatment requires a defense. Hume called justice "the cautious, jealous virtue" ([22], p. 16). Freud attempted to give this view an empirical account, arguing that the social feelings of justice and the demand for equality of treatment are derived from envy and jealousy ([20], p. 51ff; [36], p. 539). Resolution of potential or actual conflicts seems presupposed by conceptions of justice, suggesting another difference between it and the ethics of friendship or compassion.

Just arrangements, when resources are scarce, are typically complex and calculated. Often tied to specific notions about what is scarce, valued, known, believed, required or permitted, these judgments frequently lack the immediate confidence of feeling that often guides and instructs compassion or friendship. Judgments by particular societies about how to divide their finite resources and distribute goods and services differ and some, but not all of these differences are reasonable. Societies valuing sheer survival may spend more on national defense than those that do not; or in refusing to distinguish between living and living well they might choose to neglect funding for the arts, schools, museums and universities to fund intensive and chronic care facilities for permanently comatose patients.

Comparisons. Differences between actions done from friendship or compassion, and from impartiality are commonly recognized and approved; strangers do not treat us like friends or family. A *strength* of relying on compassion or friendship to direct action is that it encourages our concern for each other, not as representatives of some group, but as individuals. In doing this it moves away from a cautious and minimalistic position about what we strictly owe each other, and, if it does not harm or violate the rights of others, promotes good will between people. It encourages us to be more generous, caring and merciful. It finds a place in moral theory for meritorious and virtuous actions even by those with very limited intellectual gifts who sometimes teach us with "acts from the heart." Hume even calls these the "genuine virtues" ([20], p. 156). More recently, Simone Weil writes that justice is created from our attention to and concern for others, and "compassion sees rights from inside" ([50], p. 355).

The *weakness* of relying upon compassion or friendship alone in the formation of social policy is that it can degenerate into either inattention or promoting special interests. The needs and rights of others may be unjustly set aside for those who happen to be our friends or family or those who move our feelings. Furthermore, it can be dangerous to depend upon good will alone to look after others. It can lead to well-meaning blunders or lapses in attention to those in need. Rights protect individuals by setting minimum standards. When these rights are not adopted or clarified or have been set aside in favor of a reliance on friendship and good will, many individuals can suffer [38, 39].

To seek a just system is to presuppose that conflicts will arise that must be settled in a generally defensible way. Hume argued that systems of justice assume conditions of moderate scarcity; when there is no scarcity such systems are not needed and when the scarcity is extreme, they are ignored [22]. In setting polices, and balancing needs, wants and resources fairly and impartially, such judgments and standards may be cautious and complex. They may appear to be unsympathetic to the plight of particular individuals, to lack generosity of spirit, and to thwart rather than promote good will between us. But if duties, minimum standards of care, and eligibility requirements can offer more altruism and protection than heartfelt, spontaneous, but uninformed actions aimed at protection or altruism, then a good heart may be revealed in both simple acts of compassion and also in a firm commitment to seeking justifiable standards. Thus some justice, altruism and benevolence may be done at a distance. Compassion, however, requires attention to an individual for whom one feels something personally. Caring for *a patient* cannot be done from a distance, and the Hippocratic tradition under discussion directs that it be done empathetically rather than impersonally.

Justice and compassion generally are mutually supportive and socially beneficial. On the one hand, compassionate acts promote attention to individuals, discretion, and sympathy. They can also inspire social reform, such as trying to change for all bad laws affecting those we love. On the other hand, an ordered society committed to just treatment of citizens gives opportunities to help or attend to others and pursue friendships.

An Example of Confusing These Frameworks. There is a danger, however, in confusing what, out of compassion, we *want* for people, and what *can* be provided to all as a matter of justice. This kind of confusion may be illustrated by a common response to the question: How ought decisions for sick but incompetent people be made? The most common answer from the theorists in medicine, law, medical ethics, religion and nursing is in line with the Hippocratic tradition and the ethics of compassion. Namely, we should find out and do what is for the good of the patient or in the patient's best interest. However, we must add "all other things being equal" to the rule to acknowledge that outcomes are affected by individual circumstances and tragic dilemmas of cost, scarcity, or rationing.

To suppose, however, this means we have an obligation to provide everything irrespective of cost, scarcity or condition, confuses what we may want with what we can reasonably provide. This confusion may be found in a 1984 ruling from the Department of Health and Human Services entitled "Non-Discrimination on the Basis of Handicap; Procedures and Guidelines Relating to Health Care for Handicapped Infants" [48].[6] Let us assume its motive was compassionate: providing care to all who are in need and protecting handicapped citizens. It took the "best interest of the person" standard as an absolute rule requiring full, customary medical treatment for all but those who are dying or irreversibly comatose. The implausibility of the absolutist view presupposed by the ruling that all human life, no matter how miserable, is equally sacred has been discussed [24]; I will not pursue it further. Aside from this, the policy was bewildering and incoherent because it confused what we may want with what we can provide. Missing are cautious, calculating judgments about how this policy relates to important issues of suffering, prognosis, scarcity, resources, consent, confidentiality, or future help for these children.

Given our formal criteria of a just system, no system is just that creates unrealistic expectations or discourages informed people of good will who must use it. Those critical of this DHHS ruling found it unjust on the following grounds: The best interests of sick infants were not served by it and those caring for infants could not live with it. The government might help children more by other means than using money to police hospitals and physicians. After all, legal advances for children in this country are closely associated with pediatricians serving as their advocates. The rule ignored the traditional role of consent, confidentiality and parents as the child's primary advocates and failed to recognise consequences of the continuing medical care it requires. It said nothing about who pays and shows no awareness that often facilities are crowded, resources limited, and personnel overworked; it failed to see the possibility that in a climate of fear about withdrawing therapy from the hopelessly ill, other infants who have a good prognosis will suffer because they cannot gain admission to any intensive care nursery.

The DHHS ruling claimed support for its position from the views expressed by the President's Commission. Unlike the President's Commission, however, it ignored that sincere commitments to very impaired and handicapped infants will be costly and must go beyond the newborn period. The President's Commission states:

> When the decision is made to give seriously ill newborns life-sustaining treatment, an obligation is created to provide the continuing care that makes a reasonable range of life choices possible ... Furthermore, to the extent that society fails to ensure that seriously ill newborns have the opportunity for an adequate level of continuing care, its moral authority to intervene of behalf of a newborn whose life is in jeopardy is compromised ([49], pp. 228-229).

Thus, if no funds are provided it would seem that on the basis of the President's Commission we can conclude that the "moral authority" of DHHS "to intervene of behalf of a newborn whose life is in jeopardy is compromised."

The rule confused what some *want* with what upon reflection can *justly be provided.* It also illustrated how well-meaning but poorly thought-out policy can discourage the informed person of good will who has to use the system. This rule, in disallowing attention to circumstances and discretion in judgment, seemed to lose the very quality that motivated it, compassion for the infants.

In many countries there is a clearer consensus than in the United States about what will be provided to people. At least you should know what to expect if you are English, 70 years old and go into kidney failure - you won't get dialysis unless you pay for it yourself. Bondy points out, however, that this way of limiting services is "alien to the American philosophy that no costs are too high if a life can be saved or suffering diminished" ([5], pp. 389-390). Alien though it may seem, the views that "no cost is too great," "every life is equally valuable," and "we decide only based on the individual's best interest" are incoherent when resources are limited or scarce.

III. IMPARTIALITY, COMPASSION AND A JUST POLICY

If a just system treats alike those similarly situated, and makes us want to be just and act justly, then a just society ought to have a coherent and reasonable way of dealing with the allocation of important but scarce goods and services, including medical care. A bad policy would be one that increases harms such as by encouraging false expectations or requiring things that cannot be done. It has been argued that the Hippocratic moral duty to try to act for the good of the sick is important as a norm for a just system both in its individual interpretation and its collective one. When applied to particular persons, this duty reminds us that help comes to individuals, not cases or kinds, and that the person helped and the one who helps both matter. It promotes equity, special concern and discretion in applying just rules. But to take this Hippocratic duty as social policy requiring "everyone to do the best for everyone," not as a *prima facie* moral norm, leads to confusion about what we *want* for individuals with what can *justly* be provided. This confusion, we argued, seems evident in some recent policies. In establishing a just policy, then, we should not confuse what we *want for people* with what can fairly be provided. In addition, we should be aware of our own tendencies to *want to understand* things in certain ways. We make errors because we see what we want to see, or see certain features and concentrate on them because they satisfy us. These predispositions may even achieve a status of a coherent story or myth functioning as a consensus or account of how we ought to act. Two such biases will be suggested that could stand in the way of framing a better health care system and could encourage us to deny the need to make allocation choices.

One bias or myth is that in our society all *genuine* needs can be met including medical care for serious illness. The slogans and manifestos of this myth are like what Bondy has called a peculiarly American view, "no costs are too high if a life can be saved or suffering diminished" ([5], pp. 389-390).

It would be good if scarcity did not exist and we could devise a system insuring that all great needs could be met. For we know that liberties do not mean very much if people are too sick, ignorant, or impoverished to use them, and we understand some illnesses are caused by social factors, like poverty, ignorance, and pollution, with which people are unable to cope as individuals. If we all should have an equal chance in our society, then we need to be healthy, or as healthy as possible in order to have equal opportunity. What makes it a myth, however, is that the problem of scarcity is denied or an incoherent response is made to it: it is assumed that everyone in genuine need can receive adequate medical care, just as everyone in genuine need can receive adequate education, food or housing up to the level of equal opportunity. Although a laudable goal, the fault is in thinking resources are not finite. I will call this the *entitlement myth*. It denies the initial and tragic problem of allocation of scarce goods and services, assuming all genuine needs can be fulfilled.

Another bias or myth is that most of our so-called allocation problems are caused by regulations and society. The claim is that even though some of us are more fortunate than others, we do best when competent people take responsibility for their actions including their own health and the health of their families. We ought, therefore, to adopt an allocation policy that encourages us to be responsible, free and autonomous. I call this the *noble savage in the land of plenty myth* to show its relation to eighteenth and nineteenth century social contractarians like Locke and Rousseau. Rousseau argued that people are by nature good and society corrupts them. He wrote, "Man is born free; and everywhere he is in chains" ([40], I, 1:2). The picture that captured these writers (true or not) was one of the New World as a world of relative plenty populated by native Americans living in harmony with each other and with their environment; anyone who could put in a day's work could do well. A central problem of this myth is the false assumption we are all either capable to work out our destinies, or part of families who look out for us.

These two myths seem to have a common source: valuing equality of opportunity. They may be compelling because each focuses on different but *important* features about liberties and entitlements. The noble savage myth stresses that it is good to understand equal opportunity as equal liberty, while the entitlement mythology stresses that it is good to understand equal opportunity as a series of entitlements to have some things provided which affect equality of opportunity. The entitlement view focuses upon our complex society where misfortunes are generally undeserved and no one person is capable of achieving much without a great cooperative effort. The noble savage mythology stresses our resourcefulness in a world of plenty, and regards all regulations suspiciously as thieves after liberties. The entitlement view, in contrast, views regulations as likely champions of the opportunity to exercise liberties.

A *noble savage* mythology suggests the following as the best policy for health care: *First,* people should be held primarily responsible for their own health care and that of their families. Sade, for example, in defending this view, focuses on illnesses that are voluntarily acquired such as lung cancer induced by smoking [41]. *Second,* whatever public monies must be spent should be spent on teaching responsibility, say, through preventive care and educational programs for the many rather than crisis care for the few. *Third,* people should be notified that if they want high cost care, they must pay for it themselves, say by fee for service, or insurance or group plans. The controlling picture is of competent people bravely and stoically taking their chances and deciding how they want to face the plentiful but challenging world.

In contrast to the noble savage view that focuses on illnesses we bring upon ourselves, those favoring the entitlement myth concentrate on illnesses that are caused by social factors like poverty, ignorance or pollution, where people have little individual control. They argue that sick people are not to blame for most illnesses. The sick are harmed once by illness, again by being blamed for causing their own illness, and yet again by social indifference to their needs. It is as irresponsible to ignore the needs of the sick who need care as the needs of the uneducated and poor. Without help they are punished again for the very thing that has held them back. They are not noble savages in a land of plenty but victims suffering in an extremely complex society where help is rescue from society's evils.

These views contain important insights. On the one hand, regulations can chip away at liberties and create muddles (the DHHS regulation discussed in the last section illustrates this). On the other hand, entitlements do create opportunities. A chronically sick person cannot compete or get an equal chance. Thus, although both myths have important features, they are bad if they encourage us to make an automatic response, to dismiss the data about responsibility and suffering, or the need for such data; and to deny the tragic nature of the initial problem. For if there is no non-tragic system for allocation of resources that are scarce, then we must make choices that leave some who could have been helped without help.

IV. SUMMARY AND CONCLUSION

This paper examines the Hippocratic moral duty that one ought to try to act for the good of the sick in relation to certain requirements of justice, namely, that those similarly situated should be treated similarly and that informed persons of good will ought to be able to use the system. It was argued that some common criticisms of this Hippocratic norm miss what is important about this rule, and their charges fail. This norm does not, as charged, lead to support for either unjustified paternalism or unreasonable partiality once it is properly understood as a *prima facie* and imperfect duty. Rather, this kind of rule, by promoting compassion and equity, has an important role to play in a just system.

Unlike activities that can be done at an impersonal distance, empathy requires attention to an individual and putting oneself in the place of another. Care for individuals, the goal of medicine, cannot be done at a distance. Partiality, when it is the result of empathetic care, may be defensible in a just system.

There are two interpretations of the Hippocratic duty of beneficence: *individual* and *collective*. When interpreted as a concern for particular *individuals,* this moral rule encourages compassion for them, and the using of rules to help people who matter as *individuals*. Our benevoent, altruistic concern for the sick *collectively* is important but different. It is less partial but also less attentive to individuals. Thus, if we want a health care system that encourages sympathetic concern for individuals, as advocated by this rule, then we have to tolerate some partiality within the system. But if we want a system that also considers the good of all collectively, we have to set limits on partiality.

This Hippocratic moral duty, however, *does not* indicate the scope of either partiality or paternalism which will be acceptable to any particular society. A just society needs to clarify these limits. Impartiality and compassion can suggest different actions. If both are important features in a just system, and they can recommend alternative deeds, then their scope, relation to each other, or limitations should be defined. The formal requirements of justice do not clarify this, and the particular arrangement seems to be a choice whose reasonableness is contingent upon circumstances, resources and what is wanted, valued, or thought useful or possible. This does not mean all choices are justifiable. Rather, it means that contingent as well as formal requirements of justice need justification, and that how much discretion a society allows in applying rules is a contingent choice. It would be unjust either to create unrealistic expectations based upon uninformed compassion, or to construct a system so lacking in empathy, discretion or equity that informed people of good will cannot use it. In any allocation system, including critical health care, perceptions of what we want, or would like to believe, or find satisfying and comfortable, should be viewed as critically as anything else in framing policy.

The ancient Hippocratic norm operating within reasonable social limits is not guilty of paternalism or partiality as charged. Rather in a society that defines its limits this *prima facie* and imperfect duty of beneficence may promote virtue, compassion and justice. Whatever particular form it takes, a long-term successful policy seeks impartiality as well as equity and empathy. Thus, this Hippocratic tradition should have an important place in a just system.

East Carolina University School of Medicine
Greenville, North Carolina

NOTES

[1] The chief points being made here are logical and conceptual, not historical. As English points out, there is no single view of the Hippocratic tradition to be obtained since the corpus was slowly developed by many authors [13]. It is called 'Hippocratic' even though since Ludwig Edelstein's work it is believed that the Hippocratic Oath is Pythagorean in origin ([12], p 42ff). Both Edelstein and Veatch offer discussions of the historical development of the Hippocratic Oath [12] [47]. The duty to protect and that of positive benefit are here included under acting for the patient's good. In this paper I focus less on virtue or the Hippocratic duty *to be* a person of a certain kind and more on it as a duty *to act* in a certain way.

[2] We may disagree about what we believe fulfills these two conditions.

[3] The intentional feature, doing what one believes will help or trying to act in this way, is added since we cannot be certain what really is the good (instrumentally, inherently or intrinsically). The best we can do is to try. In the following I refer to this in shorter ways as acting for the patient's good or as the Hippocratic norm or rule.

[4] Feinberg ([16], p. 33) makes the distinction between strong and weak paternalism.

[5] The distinction between these "frameworks" is more of degree than kind, with the chief character of the choice as deliberative or not, heartfelt or not, impartial or not.

[6] More recent rules have begun to address some but not all of these problems.

REFERENCES

[1] American Medical Association: 1982, 'Principles of Medical Ethics', in T. L. Beauchamp and L. Walters (eds.), *Contemporary Issues in Bioethics,* 2nd Edition, Wadsworth Publishing Company, Belmont, California, p. 122.

[2] Aristotle: *Nicomachean Ethics.*

[3] Benfield, D. G., Leib, S. A., and Vollman, J. H.: 1978, 'Grief Response of Parents to Neonatal Death and Parental Participation in Deciding Care', *Pediatrics* **62**: 2, 171-177.

[4] Benn, S. I.: 1967, 'Justice', in P. Edwards (ed.), *The Encyclopedia of Philosophy,* Vol. 1, Macmillan, New York, pp. 298-300.

[5] Bondy, P. K.: 1981, 'Medical Care Insurance vs. A National Health Service: Impact on the Patterns of Medical Practice', *Clinical Research* **29,** 389-395.

[6] *Canterbury v. Spence,* 464 F 2d, 772, (D.C.C.:, 1972).
[7] Cooke, R. E.: 1972, 'Whose Suffering?' *J. Pediatrics* **80,** 906-908.
[8] Chisholm, R.: 1958, 'Sentences About Believing', in H. Feigl *et al.* (eds.), *Minnesota Studies in the Philosophy of Science,* Vol. III. pp. 510-520.
[9] Duff, R. S. and Campbell, A. G. M.: 1976, 'On Deciding the Care of Severely Handicapped or Dying Persons: With Particular Reference to Infants', *Pediatrics* **57,** 487-492.
[10] Duff, R. S. and Campbell, A. G. M.: 1973, 'Moral and Ethical Dilemmas in the Special-Care Nursery', *New England Journal of Medicine* **289,** 890-894.
[11] Duff, R. S.: 1981, 'Counseling Families and Deciding Care of Severely Defective Children: A Way of Coping with Medical Vietnam', *Pediatrics* **67**: 3, 315-320.
[12] Edelstein, L.: 1967, *Ancient Medicine,* edited by O. Temkin and C. L. Temkin, Johns Hopkins Press, Baltimore.
[13] English, P.: 1985, 'Commentary on Stanley J. Reiser's 'Critical Care in an Historical Context', in this volume, pp. 225-230.
[14] Evtushenko, E. 1983, 'Empathy: A Rare Gift', L. Beraha (tr.), Forward to A. Platanov, *Fierce, Fine World.* Compiled by M. Platonov, Radrega Publishers, Moskow, pp. 7-11.
[15] Fein, R.: 1982, 'What is Wrong with the Language of Medicine', *New England Journal of Medicine* **306,** 863-64.
[16] Feinberg, J.: 1973, *Social Philosophy,* Prentice-Hall, Englewood Cliffs, New Jersey.
[17] Fost, N.: 1981, 'Counseling Families Who Have a Child with a Severe Congenital Anomaly', *Pediatrics* **67**: 321-234.
[18] Frader, J. E.: 1979, 'Difficulties in Providing Intensive Care', *Pediatrics* **64**: 1, 10-11.
[19] Freeman, J. M.:1972, 'Is There a Right to Die - Quickly?' *J. Pediatrics,* **80,** 904-905.
[20] Freud, S.: 1959, *Group Psychology and the Analysis of the Ego,* J. Strackey (tr.) The Hogarth Press, London.
[21] 'Hippocratic Oath', reprinted in S. J. Reiser, A. J. Dyck, and W. J. Curran, *Ethics in Medicine,* M.I.T. Press, Cambridge and London 1977, p. 121.
[22] Hume, D.: 1751, *An Enquiry Concerning the Principles of Morals.*
[23] Katz, J.: 1985, 'Can Principles Survive in Situations of Critical Care', in this volume, pp. 41-67.
[24] Katz, J.: 1978, 'Informed Consent in the Therapeutic Relationship', in W. Reich (ed.), *The Encyclopedia of Bioethics,* Vol. 2, Free Press, New York, pp. 770-778.
[25] Ladd, J.: 1976, 'Are Science and Ethics Compatible?' in H.T. Engelhardt and D. Callahan (eds.), *Science, Ethics and Medicine,* The Hastings Center, Hastings, New York, pp. 49-78.
[26] Locke, J.: 1690, *Second Treatise of Civil Government.*
[27] May, W. F.: 1983, *The Physician's Covenant: Images of the Healer in Medical Ethics,* The Westminster Press, Philadelphia.
[28] Margolis, J.: 1984, 'Applying Moral Theory to the Retarded', in L. Kopelman and J. C. Moskop (eds.), *Ethics and Mental Retardation,* D. Reidel Publ. Co., Dordrecht, Holland, pp. 19-35.
[29] Margolis, J.: 1985, 'Triage and Critical Care', in this volume, pp. 171-189.
[30] Milgram, S.: 1974, *Obedience to Authority,* Harper and Row, New York.
[31] Nozick, R.: 1974, *Anarchy, State and Utopia,* Basic Books, New York.

[32] Pellegrino, E. D.: 1985, 'Moral Choice, The Good of the Patient, and the Patient's Good', in this volume, pp. 117-138.
[33] Pellegrino, E. D.: 1979, *Humanism and the Physician,* The University of Tennessee Press, Knoxville.
[34] Pence, G. E.: 1983, 'Can Compassion Be Taught?' *Journal of Medical Ethics* **9,** 189-191.
[35] Rawls, J.: 1955, 'Two Concepts of Rules', *Philosophical Review* **64,** 3-32.
[36] Rawls, J.: 1971, *A Theory of Justice,* Harvard University Press, Cambridge.
[37] Robertson, J. A.: 1975, 'Involuntary Euthanasia of Defective Newborns: A Legal Analysis', *Stanford Law Review* **27,** 213-269.
[38] Rothman, D. J.: 1983, 'Who Speaks for the Retarded?' in L. Kopelman and J. C. Moskop (eds.), *Ethics and Mental Retardation,* D. Reidel Publ. Co., Dordrecht, Holland, pp. 223-233.
[39] Rothman, D. J. and Rothman, S. M.: 1980, 'The Conflict Over Children's Rights', *Hastings Center Report* **10:** 3, 7-10.
[40] Rousseau, J. J.: 1750, *The Social Contract.*
[41] Sade, R. M.: 1971, 'Medical Care as a Right: A Refutation', *New England Journal of Medicine* **285,** 1288-1292.
[42] Shaw, A., Randolph, J. G. and Manard, B.: 1977, 'Ethical Issues in Pediatric Surgery: A National Survey of Pediatricians and Pediatric Surgeons', *Pediatrics* **60:** 4, Part 2, 588-599.
[43] *Superintendent of Belchertown v. Saikewicz,* 370 N.E..2d, 417 (1977).
[44] Todras, I. D., Krane, D., Howell, M. C., and Shanon, D. C.: 1977, 'Pediatricians' Attitudes Affecting Decision-Making in Defective Newborns', *Pediatrics* **60:** 2, 197-201.
[45] Toulmin, S.: 1981, 'The Tyranny of Principles', *Hastings Center Report* **11:** 6, 31-39.
[46] Veatch, R. M.: 1985, 'The Ethics of Critical Care in Cross-Cultural Perspective', in this volume, pp. 191-206.
[47] Veatch, R. M.: 1981, *A Theory of Medical Ethics,* Basic Books, New York.
[48] U.S., 45 CFR Part 84, Office of the Secretary: January 12, 1984, 'Nondiscrimination on the Basis of Handicap: Procedures and Guidelines Relating to Health Care for Handicapped Infants' DHHS, *Federal Register, 1622-1653.*
[49] U.S., President's Commission for the Study of Ethical Problems in Medical and Behavioral Research: 1983, *Deciding to Forego Life Saving Treatment,* Washington, D.C.
[50] Weil, S.: 1970, *First and Last Notebooks,* R. Rees (tr.), Oxford University Press, London.
[51] Wikler, D.: 1983, 'Paternalism and Coercion', in R. Sartorius (ed.), *Paternalism,* University of Minnesota Press, Minneapolis, pp. 35-39.

JAMES M. PERRIN

CLINICAL ETHICS AND RESOURCE ALLOCATION: THE PROBLEM OF CHRONIC ILLNESS IN CHILDHOOD

The special burdens of families with severely ill children provide a vivid illustration of the problems of combining compassion with justice in resource allocation. Both should be used in framing public policy affecting them. This paper is written from two perspectives: that of a general pediatrician caring for many chronically ill children who is also a student of public policy exploring program choices for families and children at the federal, state and local levels.[1]

Dr. Kopelman's paper [6] reviews how these individual and collective needs may conflict. Physicians, in taking the Hippocratic Oath or adhering to the rule to benefit the sick, seek to care for individuals. Kopelman points out that the Hippocratic Oath is firmly linked to the kind of norm that takes compassion for the individual as central to choice. She recognizes, however, that clinicians' resolve to act in the best interest of their patients is different from paternalism or uninformed compassion and can generate conflicts [6]. She calls for greater attention to justice in the allocation of resources and pleads eloquently for applying principles of justice and compassion together in the ethics of allocation.

In this paper I will highlight these issues by focusing on the following questions regarding care for children with chronic illness: (1) What are the demands of justice in resource allocation for children with handicaps? and (2) Does the Hippocratic Oath bind clinicians to ignore resource allocation issues in the clinical setting?

I. CHRONIC ILLNESS IN CHILDHOOD: A BRIEF BACKGROUND

Perhaps 10-15% of children under age 16 have a chronic health impairment [3, 4]. About 10% of children with chronic conditions - about 1% of the total childhood population - have a severe illness commonly interfering with the ability to carry out the usual tasks of childhood: play, school, chores, peer activities. Only a few categories of chronic childhood illness are common: allergic conditions, behavioral problems, and neurodevelopmental conditions [7]. All others are relatively rare.

J. C. Moskop and L. Kopelman (eds.), Ethics and Critical Care Medicine, 105-116.

Juvenile diabetes occurs about once in a thousand children; leukemia once in ten thousand. There are clear physiological differences among chronic illnesses, and illnesses can be classified in a number of ways, such as age of onset, impact on longevity, mobility, need for frequent hospitalization, and effects on cognitive capacities. Nevertheless, chronically ill children and their families face many issues which are similar regardless of the specific condition and which in large part distinguish these children from their able-bodied contemporaries. Families face tremendous financial expenses for medical and surgical care, often poorly covered by insurance. The illness causes significant stress on family functioning, with problems in adjustment of both the affected child and often the siblings as well. Chronic illness often keeps children out of school and thus puts them at a disadvantage in educational and social development with respect to their peers.

There are several reasons why a review of public policies in this area is timely. First, there have been tremendous advances in the technological care of many childhood chronic illnesses in the past quarter century, such that many children who would previously have died young are now living, often into young adulthood and later. Second, the chronic illnesses of childhood, for example, diabetes, asthma, cystic fibrosis, and hemophilia, have received little public policy attention in the past few decades. This is in contrast to the attention given to other groups of children with handicapping conditions, especially children with developmental disabilities and mental retardation. Finally, care for chronically ill children commands a sizeable proportion of the total dollars spent on child health in America.

II. CLINICAL ETHICS IN CHRONIC CHILDHOOD ILLNESS

Within the realm of clinical decision making with chronically ill children, two main classes of decisions may be distinguished: the severe, life-death questions and the less dramatic daily decisions regarding chronic illness. Much literature in the past decade regarding severely ill children has focused on the extreme and dramatic questions of life and death, including those of when to use respirators. The emphasis on extreme conditions with less attention to routine choices may reflect the education of physicians. Clinical training emphasizes acute illness episodes and acute exacerbations of disease rather than longitudinal experience with chronically ill or disabled people. The extreme condition has been given far greater prominence than any other issue by biomedical ethics in recent years. The ethical debate has concentrated on the circumstances under which heroic measures to keep alive a profoundly handicapped child may justly be stopped and the child allowed to die [1, 2]. For example, what should be done in the case of a child born with a significant malformation of the nervous system incompatible with higher brain functions? At what point is the care in the nursery for the 1,000 gram infant with multiple complications to be terminated? Although this debate has been clearly worthwhile, it does not address daily problems of far greater consequence to more children and families, and it often ignores the fact that "heroic" decisions take place in the complex context of ongoing family life. Questions of death and dying are rare in pediatric practice, but ethical questions involving the well-being of children and families are a daily challenge. For example, should a child be operated on when there is professional difference of opinion about probable outcomes? Under what circumstances should the activities of a child be limited when the limitations may be as damaging to the child's normal development as would be the condition they are designed to ameliorate? Who should make these decisions? At what point is the child to be brought into the decision making? A clinical example may illustrate the issues.

R.C. is a 16 year old girl from rural Tennessee, about 2-3 hours drive from Nashville. She is the first child of five in her family. Her first year of life was complicated by recurrent bronchitis and by a transient left-sided weakness shortly after her first polio immunization. At about two years of age, because of frequent infections, serum gamma globulins were obtained and noted to be low. She received intermittent injections of gamma globulin over the next few years, during which time she also had frequent ear infections and pneumonia.

At age 6, she was referred to a specialty medical center, where further evaluation supported the diagnosis of Bruton's type agammaglobulinemia, a rare condition. She was seen occasionally at the center over the next several years, though not by any specific physician. Initial entry to school was delayed because of her immune deficiency, though by 1977 she was in second grade. In 1977, a severe bilateral hearing deficit was noted, and a hearing aid prescribed. The hearing aid has been worn rarely. Developmental evaluation at that time recommended special education or a home-bound teacher. Special education services are limited in her very rural county.

In 1978, she was admitted to the hospital for pneumonia. A similar admission occurred in 1980, after which she began receiving weekly infusions of fresh frozen plasma (from maternal uncle donors). In December 1980, she developed arthritis of the wrist which progressed to other joints over the next several months. The arthritis was destructive and not responsive to aspirin, steroid, or gold therapy. She required increasing pain medication and her behavior deteriorated such that she became a difficult management problem for hospital staff and for her mother. She also developed subcutaneous abscesses, none of which grew organisms on usual culture media. Some abscesses appeared to extend into bone, and she was treated as if she had infection in the bone. She was frequently hospitalized in 1981 and 1982, for stays of about three months per year. Her mother stayed with her during almost all of her hospitalizations.

In the fall in 1982, material from some of her abscesses grew an unusual organism on special cultures. She was begun on tetracycline and has had a marked diminution in her symptoms since then. Her mood is much improved, she requires only aspirin for pain, her appetite is excellent. She has not been hospitalized again. At present, she is wheelchair-bound, deaf, developmentally delayed, and not in school.

Several questions may be asked about this example. At the time of this child's recurrent joint and abscess problems, when there was certainly little clarity as to what therapy would be in the best interest of the child, to what extent should the mother have been involved in decisions about when to use antibiotics and when not to? How does one involve this 16 year old who is developmentally retarded and functioning at the level of about a seven year old? What should be the clinician's recommendations to this mother regarding her relationship with her other children who are unaffected by chronic illness, but who are relatively neglected by their mother's attention to this child many miles from home? What options exist for educational planning for this child?

The frequent, complex decisions made in situations like this one go well beyond questions of life support, but are all too often the ethics of daily life for chronically ill children and their families. This case is further complicated by its financing. The family has limited resources. Medical services have been covered mainly by Medicaid and by the Crippled Children's Services. In Tennessee, Medicaid hospitalization is limited to fourteen days per calendar year, and the Crippled Children's Services may pay for an additional twenty days. Yet in the years 1981 and 1982, this young lady spent at least a quarter of her life in an expensive tertiary care hospital.

In addition to being this young lady's physician, I have been a consultant to the state health department on the organization of the Crippled Children's Services in Tennessee. Thus, I am aware that this is a program with severe financial limitations which, without careful restraint, will soon run out of money, leaving many children without any coverage at all. Yet, the Hippocratic Oath seems to bind clinicians to ignore resource allocation in the clinical setting. The clinical choices are several, including at least: (1) to ignore the public resource allocation issue in the care of the individual child, (2) to be judicious and restrained in recommending the use of services (including hospitalization), thus preserving some Crippled Children's Service dollars for other children (some of whom may also be my patients), or (3) to use up my share of the Crippled Children's Service funds quickly, recognizing that if I don't, my medical and surgical colleagues will consume the available dollars instead. Clinicians will increasingly face similar questions. Insofar as there is good evidence of unnecessary and at times dangerous overuse of technological efforts, there is surely some advantage to considering limiting interventions.

A different underlying question is whether clinicians should have to take into account the problems of limited resources when dealing with specific patients in their practice. One non-medical colleague asked if potential consumers should know whether a physician considers resource allocation in clinical recommendations, suggesting that patients would like to know and would likely choose physicians offering "the best" without concern for limited resources.

Another area illustrates the way in which clinical problems or even successes become a matter for public policy. The past decade has seen the emergence of a new class of pediatric specialist, the pediatric pulmonologist or lung specialist. With increasing sophistication in the management of lung diseases and, more importantly, the maintenance of lung function in the face of other severe conditions, many children are surviving today with major respiratory support, almost all of whom would have died a decade ago. About a third of these children are survivors of intensive care nurseries, a third are the result of major trauma, usually motor vehicle accidents causing paralysis by transecting the spinal cord at a high level, and about a third are a miscellaneous category including children with muscular dystrophy. These children, perhaps now three hundred or more nationwide, live in the intensive care units of America, often spending years there. They are so highly dependent upon equipment and excellent pulmonary nursing that they cannot be moved to regular wards. Pediatric intensive care units (noisy, sterile, and complex environments, usually lit twenty-four hours a day) are not conducive to the child's best development.

Conservatively, the yearly cost of care for these children is about $150,000 [8]. Recently, a few centers have experimented with developing smaller, mobile pulmonary units, permitting children with the best function to leave the hospital and be cared for at home with their equipment trailing around behind them. It is likely that providing services at home will cost less, though still on the order of $80-90,000 per year. Most respirator dependent children have run out of their insurance coverage. In reality, their hospital costs are paid mainly by increasing the daily hospital rates of other hospitalized patients. Such cross-subsidization is not possible for care at home, and there has recently been a call for public support for such children.

Again the clinical and resource allocation questions come together. Should the specialist caring for these children take into consideration the tremendous costs and limitations of resources in providing excellent pulmonary care for them? Will the institution of a national, publicly supported program for home care have an impact on the number of children who are respirator-dependent? Who won't get services if these children do? Can clinicians reasonably expect public support when the results of their efforts are very costly?

III. JUSTICE IN HEALTH RESOURCE ALLOCATION

Health care for chronically ill children is financed through several mechanisms, including private insurance, Medicaid, voluntary associations, research grants, and the Crippled Children's Service. The Crippled Children's Service illustrates well the issues of health resource allocation. Among the oldest of federal grant-in-aid programs (begun over fifty years ago), the Crippled Children's Service has always been a state option, states' rights program. Unlike later more centralized health initiatives, individual states have always implemented and developed programs and policy. From a policy perspective, this has been a prototypic block-grant program. The original legislation emphasized "crippled children" in the name of the program, but defined crippled children very broadly, including most children with chronic illnesses as we now think about them and excluding only psychiatric disorders, "incurable" disorders, and mental retardation. Although the program was initially open to all eligible children without means testing, in essentially every state, income has become a major determinant of eligibility. Indeed, despite the great variability in programs among states, perhaps the only consistent characteristic among Crippled Children's Service programs is the presence of means testing. The program has been implemented in a number of different models, varying from serving mainly as an insurance program providing little direct service (but rather insuring the population just above Medicaid eligibility), through contracting for services (often from academic health centers), to the direct provision of services by Crippled Children's Service clinics. The program generally emphasizes bringing available medical and surgical therapeutics to children with chronic conditions. When new technologies develop, the program changes to incorporate them. Thus the program has expanded tremendously from its original concern with treatable orthopedic conditions to operable congenital heart disease to now many pediatric medical diagnoses. In most states, it is now a chronic illness program rather than a crippled child program.

Though an important source of financing for some children with chronic illnesses, the program has never been adequately funded, and beginning in the 60's it has been far out-shadowed by Medicaid as the main source of financing for health services to chronically ill children. In the past two years, the program has suffered a major decrease in federal support of about 30%, which has been buffered at the state level by sizeable state contributions. The resulting adjustments in programs illustrate the problems of health resource allocation.

Resource allocation decisions have rarely been explicitly discussed, either in times of program expansion or now in times of service cutbacks. Indeed, there may be some political risk in sharing decision making broadly or making the criteria explicit. How decisions are made has not been well studied. State Crippled Children's Service directors were recently asked with whom they would consult to determine reduction in services and what would be the specific issues in doing so. For the most part, the directors turned for guidance to other state administrators and then to advisory groups made up largely of physicians. Only thirteen state directors reported that they occasionally consulted with parents or community groups or voluntary organizations; no state reported frequent or regular consultation with such groups. Should there be public scrutiny of these decisions? How are the concerns of consumers considered?

What criteria may be used to decide who gets services? Several broad approaches to justice in the distribution of health resources are available.

To each according to his merit. The notion of awarding goods on the basis of merit is familiar in our society. To the extent that the possession of money is an indicator of merit, however imperfect, merit plays an important part in determining who gets what services and what kinds of services they get. While there is wide acceptance of this principle in general, no one seems to be completely happy with its application in matters of health or medical care. Unlike automobiles, vacations, and housing, health or medical care appears to have a prior claim on resources and to be more like a human right. As a means for allocating resources to children, especially for those with chronic and severe illness, methods according to merit seem hard to justify.

To each according to his societal contribution. More health resources might be allocated to those who contribute most to society. To some degree, in the early days of dialysis, committees deciding on the allocation of scarce resources did judge the societal contributions of candidates for dialysis, though the problem of assigning relative values to varied contributions made the process unmanageable. For children, the problem is even more tangled. Here we are considering likely future societal contributions. In this situation, one might allocate greater resources to a child with diabetes because his intellectual capacity is probably greater than that of a child with a myelomeningocele.

To each according to his need. If health and medical care are human rights, then it may make sense to allocate resources based on some measure of need. Operationalizing need and distinguishing it from demand or desire are often quite difficult. With scarce resources, there are never enough to meet all health needs. There then remains a necessity of achieving agreement on which is the greater and the lesser need in order to fix priorities.

To each on a random basis. Much of the present allocation of resources, especially in hospital settings, is indeed based on a first come-first served pattern. While not ideal nor random in the sense of a lottery, if one can assure equality of access to a limited resource, then perhaps this kind of random allocation (first come-first served) is as just as any other.

How are these concepts of allocation applied in reality and what are the choices that program policy makers face? The decisions are complex, and the criteria upon which decisions can be made are at best confusing and conflicting. In broad terms, there are perhaps three ways in which administrators can decide to limit services or in other times to broaden them. The *first* is to change eligibility for services based on such criteria as age or income, the *second* is to limit for all the types of services provided, and the *third* is to exclude children on the basis of specific diagnoses. Let us consider each of these in turn.

(1) In general, criteria which reflect population characteristics are easier to administer than are ones which require judgments about the need for specialized services. Population criteria, which reflect family or demographic variables, include age, socio-economic status, and

occasionally geography. In administering the Crippled Children's Service program, one can change the age of eligibility from 18 to 15 or the income for a family of four from $8,000 a year to $6,000 a year. Making people be poorer before they can receive Crippled Children's support makes little sense in considering the needs of chronically ill children and their families, whether or not it makes sense to have separate private and public health care systems for people in general. When considering individually relatively rare phenomena like the appearance of a chronic illness in a child and when there is a need for often very specialized and relatively scarce resources, chronic illness programs for the public sector separate from those for the private sector tend to diminish access and lead to costly duplication of services.

(2) The needs of families with chronically ill children are complex. They include access to high quality specialty services, as well as high quality general health services, which they need perhaps even more than do families with able-bodied children. Other services frequently needed include home nursing care, psychological counseling, genetic counseling, and aid with the tremendous financial burdens of chronic illness. What range of services should public programs offer? Does it make sense to provide mainly medical and surgical services to a family with a child with spina bifida, and not to provide genetic services, which might affect the reproductive activities of the family? The goal of public programs should be optimal functional outcomes for children, given the family and society's resources. Yet, focus on the medical and surgical needs of a child with, say, muscular dystrophy may mean that his knees will have better mechanical function, but that he may never attend school after the age of eight. A child with spina bifida who has technically excellent orthopedic surgery but no follow-up or inadequate physical therapy or inattention to working with his school to make necessary adjustments to his environment or to the provision of needed school health services will perhaps be able to move his joints but may not learn or become a productive citizen. The goals of our care and work with families should be to maximize the family's capacity to meet its own needs and the child's ability to participate as much as possible in usual daily activities for his age: school, sports, clubs, dating, and jobs. Providing medical or surgical care, though necessary, is not sufficient to lead to these outcomes. Thus, despite the attractiveness of limiting services to physician-based care or hospital care, public programs need to develop a minimum set of services, the provision of which is likely to improve broad family outcomes.

(3) If neither limitation by demographic variables nor by scope of services seems appropriate, does limiting the conditions to be covered make sense? This approach too offers a variety of questions and could be implemented in several ways.

(a) Should resources be limited to children for whom there is a good prognosis? Is curability or repairability a reasonable criterion, that is, serving first those children who can be restored to complete or near complete functioning? This policy places as a lower priority children for whom treatments may improve functioning, but have no likelihood of restoration to full capacity.

(b) The availabilty of specific procedures to treat a condition may be a driving force in deciding the conditions covered. The early history of the Crippled Children's Service reflects an initial availability of orthopedic procedures and an expansion into cardiovascular conditions with the onset of rheumatic fever prevention and the surgical treatment of congenital heart disease. Yet the fact that the condition is treatable, that is, that it has a procedure which can be applied to it, does not assure that the procedure will improve the health or functioning of children in any or all circumstances. Should resources be allocated because there is a surgical procedure or medical treatment available or because it is known that the procedure or treatment improves the patient? Yet, as many therapies are untried and unproven (but frequently used), patient improvement may be hard to document.

(c) Should services be limited to high cost conditions, recognising the catastrophic financial impact of chronic illness and allocating resources where the need is greatest? Children with more limited or less intensive or less expensive problems may rely on their own resources. This policy will select high cost conditions such as spina bifida or congenital heart disease and exclude such chronic medical conditions as diabetes and asthma. The catastrophic criterion may conflict with a policy to maximize improvement in functional outcome among children with chronic conditions for the number of dollars spent. Allocating resources to many children with low cost conditions, for example, might lead to greater overall benefit than might focusing resources on relatively few children with catastrophic conditions.

(d) Finally, at a time of cut-back, one might simply decide to remove conditions which have recently been added to the program rolls. In Ohio, services for the prevention and treatment of genetic conditions have been removed from conditions covered, mainly because they were added recently. An even more pragmatic approach may arise from the assessment of the relative political strength of interest groups supporting one condition or another. The preservation of political support for the program might dictate that, if the local cystic fibrosis foundation and support groups are stronger than those for diabetes or sickle cell anemia, cystic fibrosis would receive a higher priority in that state's allocation.

VI. SUMMARY

If the clinician's ethical stand is to do what is in the best interest of the patient and family, none of the policy options above is terribly appealing. Yet these are indeed the decisions faced on a daily basis by policy makers. In general, the same issues can be applied to a wide variety of public programs, not limited to health or to the health of children with chronic conditions. One of the institutional methods of separating the ethics of compassion from the ethics of justice has been to place most allocation decisions in the public sector, with direct health services provided mainly in the private sector. Thus one has compassionate friends (physicians) delivering services to chronically ill children with impartial administrators deciding which patients actually receive coverage. Yet the fundamental conflict is between assuming that no limitation of care is justified and recognizing that there are (and will continue to be) marked limitations of resources. Limits of care exist now, and the ethical analysis may best proceed by describing the values reflected in present day resource allocation decisions.

It does seem that the tension between the ethics of friendship and the ethics of justice may be productive. The clinical approach to patients is often talked of as a special way of problem definition and resolution, perhaps a special world view of clinicians trained in any of a number of disciplines. Yet the clinical approach is also a special type of compassionate experience. Clinicians on the other hand have little experience and exposure to matters of resource allocation or the ethics of justice. Their special form of compassion, linked with a better sense of what may be due to all similarly situated as a matter of justice, may let clinicians participate more effectively in the solution of health problems. Furthermore, clinicians will become increasingly aware of issues of resource decision making in the context of direct care of their own patients. How much tension (between compassion and justice) can clinicians tolerate? What methods can be implemented to assure attention to these hard decisions? Though the quesions are never easy, they will be answered better if we attend to them carefully. Hopefully the ethical questions in medical care can expand well beyond the ethics of the individual doctor-patient relationship to the serious considerations of the ethics of public policy.

Vanderbilt University School of Medicine
and Institute for Public Policy Studies
Nashville, Tennessee

NOTE

[1]I am deeply indebted to my late colleague Professor Nicholas Hobbs, who felt strongly that a study of values is essential to the policy analysis process, even though the explicit statement of values may be anathema to policy makers and politicians. The author wishes to thank Drs. Virginia Abernethy, Mark Sullivan, and Richard Zaner for their insightful comments on an earlier version of this paper. This paper includes material from a chapter of *The Constant Shadow: Childhood Chronic Illness in America* [5].

BIBLIOGRAPHY

[1] Duff, R. S.: 1981, 'Counseling Families and Deciding Care of Severely Defective Children', *Pediatrics* **67,** 315-320.
[2] Fost, N.: 1981, 'Counseling Families of Children with Severe Congenital Anomaly', *Pediatrics* **67,** 321-324.
[3] Gortmaker, S. L.: 1985, 'Chronic Childhood Diseases: Demographic Considerations for Public Policy', in N. Hobbs and J. Perrin (eds.), *Chronically Ill Children: A Stacked Deck,* Jossey-Bass, San Francisco.
[4] Ireys, H. T.: 1981, 'Health Care for Chronically Disabled Children and Their Families', in *Better Health for Our Children,* report of the Select Panel on the Promotion of Child Health, Vol. IV, U.S. Govt. Printing Office, Washington, D.C., pp. 321-353.
[5] Hobbs, N., Perrin, J. M. and Ireys, H. T.: 1985, *The Constant Shadow: Childhood Chronic Illness in America,* Jossey-Bass, San Francisco.
[6] Kopelman, L.: 1985, 'Justice and the Hippocratic Tradition of Acting for the Good Of the Sick', in this volume, pp. 79-103.
[7] Pless, I. B. and Roghmann, K.: 1971, 'Chronic Illness and Its Consequences: Some Observations Based on Three Epidemiological Surveys', *J. Pediatrics* **79,** 351-356.
[8] U.S. DHHS: 1983, *Report on the Surgeon General's Workshop on Children With Handicaps and Their Families,* DHHS Publication No. PHS-83-50194, Washington, D.C.

EDMUND D. PELLEGRINO

MORAL CHOICE, THE GOOD OF THE PATIENT, AND THE PATIENT'S GOOD

I. INTRODUCTION

Acting for the good of the patient is the most ancient and universally acknowledged principle of medical ethics. It grounds ethical theories and shapes the way their principles are applied in particular cases. It is the ultimate court of appeal for the morality of medical acts. While it may, on rare occasions be set aside for the common good, this is done with trepidation and in only the most urgent circumstances.

Yet what precisely we may mean by the patient's 'good' or the 'good of the patient' is subject to the most different interpretations. These divergent interpretations engender some of the most vexing ethical dilemmas, and their solution is impossible without clear understandings of the central terms. But so beguiling is the idea of doing good for the patient that we rarely examine closely what good, and whose good, we are serving.

In a morally diverse society, opposing views of ultimate and immediate good may be held by the parties in clinical decisions involving moral choice. Each participant is a moral agent and as such is bound to uphold, and be accountable for, his, or her, own conception of what is right and good. Making morally defensible decisions in the face of substantive differences in conceptions of patient good has become, therefore, one of the urgent procedural problems in medical ethics.

The problem is, of course, a subset of the problems attendant on the lack of moral consensus in our society. This, in turn, is the consequence of our philosophical and theological disagreement on what constitutes "The Good," and the good life. A theory of good grounds every theory of morals, general and medical. Since we are not likely to agree on our philosophical or theological definitions we are compelled to clarify the various senses in which we may use our terms and establish a morally defensible procedure for dealing with conflicts when they arise.

This essay is an attempt to meet these two requirements - an analysis of the components of patient good and a procedure for handling differences in a morally defensible way. It proposes that the concept of

J. C. Moskop and L. Kopelman (eds.), Ethics and Critical Care Medicine, 117-138.

patient good is a compound one, that at least four senses of patient good can be discerned, that they are related to each other, but distinct, that the physician (and other health professionals) are obliged to respect each level of patient good, and that a hierarchy exists among them that determines how conflicts should be resolved. Moreover, each component can be related to the several notions of the good in Aristotle's *Ethics*.

The author's analysis of patient good, together with related essays on the philosophy of the healing relationship and the virtuous physician, constitute an effort to complement and supplement the prevalent emphasis on rights and duty-based ethical systems [23].

Because of their urgency, and their momentous nature, decisions not to resuscitate - so called "No-Code Orders" - will be used to illustrate how different notions of patient good may conflict and why a clarification of the meanings of patient good is necessary. This is so even with the benefit of the recent, admirably cogent analysis of life-sustaining treatments by the President's Commission [6]. That report clearly recognizes the centrality of patient self-determination as well as patient benefit in the decisions to refuse cardiopulmonary resuscitation and other life-sustaining measures. Even with such well-reasoned guidelines, the question of what is the good of the competent and the incompetent patient remains at the heart of each decision.

II. THE PATIENT'S "GOOD": ITS FOURFOLD MEANINGS

The good of the patient is a particular kind of good, that which pertains to a human person in a particular existential circumstance - being ill, and needing the help of others to be restored, or to cope with the assault of illness. In a general way the good the patient seeks is restoration of health - a return to his or her definition of what constitutes a worthwhile way of life - one that permits the pursuit of personal goals with a minimum of pain, discomfort, or disability. This is the end the patient seeks in the medical encounter, and the physician promises to serve by his act of "profession" - his promise to help with the special knowledge at his disposal. The physician thus becomes an instrument for the attainment of the good the patient seeks.

Inherent in the nature of the physician's offer to help is a tacit promise to use his knowledge and skill to advance the patient's good [10, 22]. This may be interpreted differently by the patient, the physician or the family. Unless the patient is incompetent, however, the physician is obligated to act for the good conceived by the patient and to support his goals. In the event those goals are morally unpalatable to the physician, he is free to withdraw from the case under the usual conditions.

If he accepts the case, and as long as he maintains his relationship with the patient, the physician is obligated to promote four components of the patient's good: (1) *Ultimate good* - that which constitutes the patient's ultimate standard for his life's choices, that which has the highest meaning for him; (2) *Biomedical or techno-medical good,* that which results from the correct application of medical knowledge and skill; (3) *The patient's perception of his own good* at the particular time and circumstance of the clinical decision and how he prefers to advance his own life plan; (4) *The good of the patient as a human person,* capable of reasoned choices. Where the patient may have confused or conflicting notions of his own good, full congruence may not be possible. Nonetheless, I will argue that the physician is bound to advance each of these four senses of good to the extent possible.

(a) *Good, the Good, and the Patient's Concept of Ultimate Good*

At the outset, and throughout this discussion, we must remain clear about the distinction between the good as perceived by the participants in clinical decisions, and the ontological nature of good. This essay cannot presume to deal adequately with the prickly question of the objectivity or non-objectivity of the good. The point of this essay is not whether particular interpretations of patient good are metaphysically sound. Rather its focus is on the fact that widely divergent interpretations do occur, that in spite of that fact, physicians, patients and families must make decisions together, and that the conflicts, when they occur, must be dealt with in a morally defensible way.

But even if all the participants in a clinical decision were to agree on each of the four levels in their interpretations of patient good, that would not make the decision ontologically good. One could conceive of agreement on interpretations that would be intrinsically evil - e.g., falsification of disability or cause of death to gain compensation, or insurance benefits; withholding public information of a diagnosis of plague or cholera, or of a patient's homicidal intent; the practice of killing defective infants or adults.

The aim of the essay is to encourage a clearer identification of each component of patient good, to clarify the use of this universal notion, to understand the conflicts that can result from its varying interpretations by the participants in clinical decisions involving moral choice, and to examine how to resolve these conflicts in some orderly, and morally reliable way.

The good of the patient is a particular good and like all particular goods it is related to, and shaped by, the conception we hold of the notion of good and "The Good." This is the first component to be examined in mapping the content of the good of the patient. What we think of the nature of good, and "The Good" ultimately shapes all the other components.

The history of philosophical debate about the nature of "the" good is too long and unsettled to be repeated here [26, 5, 19, 11]. The perennial question remains - what is the nature of "the" good? Is it properly understood as one thing or many? Does it rest on some factual aspect of the nature of man or the world? Is it primarily a psychological, intuitive or self-evident concept? Is it even susceptible of rational justification? Is the ultimate good the contemplation of truth, living in accordance with God's will, developing one's potential, wealth, honor, pleasure, power, the good of the species, or some combination of these things that adds up to "happiness"? We need not resolve these questions to appreciate that how we answer conditions the subsidiary notions contained in the ideas of the patient's good.

The concept we hold of the ultimate good is the reference standard for all decisions including clinical decisions. It serves to justify and define the nature and aim of moral choices. In clinical decisions, some take the ultimate standard to be whatever the patient desires, others what the physician judges to be good, for still others conformity with philosophical, theological, or socially determined principles like the will or law of God, or freedom, self-determination, social utility, quality of life, or the good of the species. The list is long, and the competing concepts are often incompatible.

Equally incompatible are the opposing theories of those who hold that the good is whatever humans desire, or have an interest in, and those who hold the good to inhere in certain things and actions whether humans desire them or not [6]. Without trying to resolve these theoretical questions, it is sufficient to realize that in any particular moral choice some final concept of the good, some "good of last resort," underlies the other components of patient good. That final concept is the most pervasive, the least negotiable and often the least explicit presupposition when conflicts arise in making clinical decisions with moral overtones.

(b) *Biomedical or Techno-medical Good: What Medicine Can Achieve Technically*

Biomedical or techno-medical good encompasses the effects of medical interventions on the natural history of the disease being treated. It is the good that can be achieved by the application of expert technical medical knowledge - cure, containment of disease, prevention, amelioration of symptoms, or prolongation of life. It is directly related to the physician's technical competence; it is the first step in fulfillment of the moral obligations of his or her promise to help. Biomedical good is the *instrumental* good the patient seeks from the physician. It is also a good internal to medicine - part of its claim to be a special kind of human activity. It is the good that results from the physician's craftsmanship - his capacity to make the technically correct decision and to carry it out safely, competently, and with minimal discomfort to the patient. Biomedical or techno-medical good is usually subsumed under the phrase 'medically indicated'.

There always is an unfortunate tendency for physicians to equate biomedical or techno-medical good with the whole of the patient's good. Techno-medical good does not exhaust the good the physician is obliged to do. It is an essential but not a sufficient component of good medicine. Two ethical errors may result from the conflation of techno-medical good with the good of the patient.

The first error is to make the patient a victim of the medical imperative to insist that if a procedure offers any physiologial or therapeutic benefit then it must be done. On this view, ethical medicine is limited to technically right interventions. Ethical quandaries are thus ignored since the only good acknowledged is medical good in its narrowest sense, and this is ascertained by scientific means and not by ethical discourse or analysis.

The second error is to confuse the physician's judgment of the tolerability of the quality of life that would ensue from a treatment with the medical indications for that treatment. On this view, if treatment of a defective infant results in a life without "meaningful relationships," then it is not medically indicated and should not be done. This is an unjustifiable extension of medical judgment beyond its legitimate limits. Whether a life is worth living is a value decision only the patient who must live that life can decide. Moreover, it is not a matter determinable by the capabilities of medicine *qua* medicine, in the first place.

It is the obligation of the physician to ascertain, by the most careful method, the kind of life that might ensue from a particular treatment in a particular patient. These are matters of scientific judgment proper to medicine. They are essential in helping the patient to decide if the life that ensues from treatment is worthwhile. It is he who must judge, with the physician's help, the kind of life he wishes to lead and the risks or discomfort he is willing to bear to attain the benefits medical treatment might offer.

On this view, I would argue that medical good but not medicine should be interpreted narrowly - as that which can be ascertained scientifically and technically to be possible in altering the natural history of *this* disease in *this* person. I recognize that certain value judgments are involved but these should be kept to a minimum and limited to scientific competence, sound clinical reasoning and valid probabilistic statements about diagnosis, prognosis and therapeutics. I must repeat here that while techno-medical good should be defined narrowly, this is *not* the whole of the physician's responsibility or of medicine as a practice.

Let us examine this concept of bio-medical good more specifically in the context of cardiopulmonary resuscitation (CPR).

CPR was introduced in 1962 to reinitiate the heart beat in patients about to die suddenly due to electrical dysfunction of the heart, in the presence of a mechanically intact heart muscle. A variety of dysrhythmias - ventricular, fibrillation, asystole or bradycardia - may lead to ineffective output of the heart with cessation of blood flow to the brain. Within a few minutes cerebral damage becomes irreversible and death ensues.

The ideal and unequivocal clinical indication, the highest techno-medical good that CPR can achieve, is in cases of sudden death resulting from acute cessation of flow through a coronary artery in an otherwise normal heart. Resuscitation offers the probability of a complete recovery and subsequent use of effective medical and surgical procedures for the underlying disorder. This is true also in drowning, electroshock, or when someone is struck by lightning. In these situations, the medical good is unequivocal.

Sudden cessation of cardiopulmonary function may also occur under other clinical circumstances in which the final medical outcome may be less clearly predictable. Cardiopulmonary arrest, for example, can accompany any severe acute illness from a variety of causes - medical and surgical shock, overwhelming infection, diseases of the brain and meninges, diabetic ketoacidosis, poisonings, or renal failure. In these situations the heart muscle may or may not be intact, but if the patient can be resuscitated, treatment can be instituted for the underlying disorder. CPR is medically indicated though the eventual good is problematic, and cannot usually be foreseen at the time CPR is instituted. The techno-medical "good" here is to gain time to diagnose and treat the underlying disease.

The most controversial use of CPR is in the patient who is known to be dying - one whose underlying disease has progressed to the point of no return, and in whom further treatment offers little or no prospect of success. Here cardiac and pulmonary arrest, and the accompanying cardiac dysrhythmias, are simply part of an inevitable process of dying. From a strictly biomedical point of view, in these situations CPR only temporarily interrupts the inevitable last stages of dying. If CPR is successful, a series of other measures will usually be required to keep the patient alive - intubation, mechanically-assisted respiration or circulation, drugs to support circulation, infusions, dialysis, etc. These measures offer a very limited medical good - to sustain life (usually for other than medical reasons) - until some decision is made to permit the process of dying to resume its inexorable course.

In cardiopulmonary arrest outside the hospital, it is rarely possible to determine accurately whether the patient will be medically assisted by CPR. CPR could achieve some techno-medical good since it gains time for the patient to be transported to the hospital where more deliberative and definitive diagnosis and prognosis are possible. Generally speaking, CPR under these circumstances would also coincide with patient good as we shall describe it shortly unless the patient had previously been diagnosed as terminally ill on good clinical grounds or has executed a valid living will. Whenever prognosis and diagnosis are uncertain, CPR can serve some medical good by allowing time for a better assessment of what medical interventions can offer but also what the patient's desires might be with respect to a no-code order.

(c) *The Patient's Best Interests: The Patient's Concept of his Own Good*

A biomedically or techno-medically good treatment is not automatically a good from the patient's point of view. It must be examined in the context of the patient's life situation and his or her value system. To be good in the fuller sense, the choice must square with what the patient thinks worthwhile given the circumstances and alternatives his illness forces upon him. The patient must weigh the probable medical benefits of a treatment or a no-code order against some ultimate good (e.g., his religious beliefs). When he is competent, only the patient can decide ultimately whether the quality of the life that remains is "worthwhile," consistent with his belief system or with some plan he may have for his life. The range of satisfactions left to the sick and disabled is narrowed. But what is left may still be savored by the patient in ways the healthy person cannot comprehend.

When the patient is competent it is he who can best ascertain what is in his best interests. When he is not, then his surrogates must ascertain as closely as possible what he would have chosen as in his best interests were he able to make the choice himself. Our concern must be for the person who is to live the life illness imposes, not what we think of the quality of that life.

The court in the Shirley Dinnerstein case [14] asserted the legal import of the distinction between the medical decision - based on diagnosis and prognosis - and the personal decision - based on the surrogate choice by the family of what was in their mother's best interests [31]. The medical good of a no-code order was deemed within the doctor's province; the patient's interests were placed in the hands of the family. Both were relevant for a legal decision not to resuscitate.

The patient's "interests" then, are the aims, plans, and preferences peculiar to him and chosen by him at a particular time. Any object of desire may become an object of interest for this patient at a particular time. The "good" in this sense can be anything which is an object of interest for this patient.

Patient interest defined in this way is necessarily subjective and relative, since it is rooted in *the patient's view* of what is in his own best interests at this time, and in this circumstance. We cannot know what that is until we ask the patient. This view does not deny the possibility that some objects of interest will be bad or injurious - it requires only that they have been freely chosen by this patient. It is no revelation that we may know the good, but do not infallibly choose it. Because someone has chosen something as good does not make it good intrinsically, or instrumentally. To accept the patient's definition of his own best interests does not necessitate that the doctor agree nor is he bound morally to promote those interests. What the physician must do is to give the most serious weight to the patient's judgment of his own interest in making decisions. Indeed, that judgment must be accorded primacy, since it arises from the operation of an even more fundamental good - the human capacity to choose.

(d) *The Good of the Patient as a Person*

The fourth sense of patient good is that which is most proper to being a human person. It is a somewhat different category from the other three senses of good which I have defined. Each of these may be individually determined for, or by, a particular person in a particular circumstance and weighted differently in different persons and circumstances. The fourth good is the operation of the capacity to use reason to *make* choices, and to communicate those choices through speech. One cherished and distinctive feature of human existence is this capacity to establish a life plan, and to select from a variety of goods those things that are preferred for reasons that are unique and personal. Humans may not reason wisely, prudently or correctly, but the freedom to do so is a good without which it is impossible for the mentally competent person to live a good life.

Those humans who by virtue of pathological or physiological abnormalities of brain function cannot make choices - the comatose or psychotic - or, those who have never been competent - infants - are still humans. Their choices must perforce be represented by others but they must be represented nonetheless. Even though their capacities to make reasoned choices cannot be expressed because of brain dysfunction, they are still beings whose nature it is to be rational. We are compelled to respect their good in this manner to the extent possible by the alternative means of surrogate or proxy decision.

Choice requires freedom, and freedom implies that some choices may be wrong or evil. Liberty and the power to choose are therefore intrinsically bound to each other. Louis Lavelle puts it well:

> For liberty is nothing if it is not the power of choosing . . . Thus the perfect unity of the self lies in the possibility it has of choosing. But it can only choose between alternatives and the self's unity is the living unity of the act which postulates and resolves this alternative . . . We can see therefore that by a kind of paradox our liberty can determine itself only by distinguishing between good and evil in the world ([16], p. 18).

If we are not to violate the humanity of the patient in medical decisions, so long as the patient is competent, we must allow him to make his own choices. We cannot override those choices even if they run counter to what we think is good for the patient. To manipulate the patient's consent, to deceive or misinform him, even to do what we think is good is to violate his good as a human being. Only the patient can free us of the obligation to abide by his choices by giving us a mandate to make decisions for him if he feels emotionally or intellectually overwhelmed by the complexity of the choices. But even in the act of yielding up his prerogative, the patient exercises his freedom because he can choose not to exercise his capacity if he wishes. The physician can never presume to usurp that prerogative. The freedom to choose, and to be responsible for the outcome of those choices, is the ground upon which any reasonable notion of autonomy is built.

The good of the patient as a human is therefore a more general good than the others. It is the basis for our respect for the personal interpretations particular patients in particular circumstances may place on techno-medical good, their immediate interests and their ultimate good. While the ultimate good is the highest good, it too must freely be chosen by the individual. Without freedom to choose and reject there would be little merit in subscribing to some ultimate good, no matter how lofty.

In every human decision, and especially in clinical decisions, these four senses of the good are intermingled. The configuration of choices we make at each of these levels and the way we relate one to the other in large part defines us as persons. Each level must be understood and respected. When they are in conflict with each other, or are interpreted differently by the parties making medical decisions, some hierarchical order must be established among them. Without such an order decision making is paralyzed or, worse still, results in the capitulation to or superimposition of one person's choices on another.

To acknowledge that these four levels of good can differ among interacting humans is not to accede to the idea that all choices are equally valid logically, epistemologically or metaphysically. No capitulation is being made here to moral relativism or emotivism. There are morally good and bad choices at each level, and those distinctions must not be abandoned. They must be recognized, and morally sound procedures established, so that each element of patient good is adequately promoted in the face of potential conflicts among the interacting parties. The ethics of the procedure must be distinguished carefully from the ethical substance of the decisions.

III. NO-CODE ORDERS AND THE COMPONENTS OF PATIENT GOOD

No-code orders illustrate some of the conflicts and confusions that may occur in clinical decisions among persons all of whom would argue that they are pursuing the good of the patient, or avoiding harm, or respecting patient autonomy [1, 2, 7, 9, 12, 15, 17, 18, 25, 27].

The most difficult situation occurs when CPR is not medically indicated because the patient's underlying disorder is incurable or the patient is in a terminal state. CPR will then only interrupt the last stages of dying; that is, no techno-medical good can be achieved in the sense that the natural history of the illness cannot be reversed.

Under these circumstances, the patient, physician, or family might still wish to use CPR because of some "interest" of the patient. For example, the patient or physician might be an Orthodox Jew who feels that the doctor's dedication must always be to preserve life at all costs - as long as something can be done it must be done. Many Christians hold this view also. On this view, life will end only when God wishes it to end. God would not have given man the knowledge of CPR if it were not to be used to preserve life. Any relaxation of efforts to sustain life would be interpreted as nothing short of assisted suicide, perhaps even homicide.

On a more secular level, CPR might be chosen, even when its benefits are dubious, to take advantage of the remote possibility of diagnostic or prognostic error. Even with the best of human knowledge prognosis can be a risky exercise in probabilities. Any clinician with extensive experience can recall patients he consigned to inevitable death who confounded him by surviving. Also, there is always the remote possibility that if the patient is kept alive some new treatment for the underlying disease might be discovered. Or, some bold therapeutic idea might occur to clinicians or consultants that might overturn the hopeless prognosis. Obviously such cases are vastly in the minority. Still, for some patients whose values dispose them to a last ditch stand, CPR might be chosen even though its utility may be virtually nonexistent.

Another example is the case in which patient, physician and family agree with the prognosis and the medical futility of CPR yet find that the patient perceives a few additional hours, days, or weeks of life to be in his interests. The patient, before he dies, may wish to see a child or grandchild born, a son graduate, a long absent relative, or friend. Or, he may wish to finish some work, or attend to some final arrangement regarding his estate, or fulfill some obligation.

CPR may also be in the family's or physician's interest (but not in the patient's); it may assuage their guilt, real or imagined, for the way they may have treated the patient during his life. If "everything" is done, they might feel some part of their past guilt to have been expiated. In this way also the physician might assure himself that he is not "responsible" for the death. Sometimes medical attendants are reluctant to let a patient die when they have invested months of work and personal involvement in his care. In these situations the patient's good is in conflict with the needs and interests of his family or medical attendants. Under the guise of serving the "good" of the patient, they may violate the patient's perception of his own good and his freedom to make his own choice as a human being.

There are also cases where the physician does not wish to use CPR even when it might be medically indicated, because he believes he would thereby "condemn" the patient to a life not worth living. To avoid "doing harm" the physician might withhold CPR, for example, in an elderly man who had a stroke or suffered an acute coronary episode. Or, the physician might decide that preserving the life of a malformed infant or a retarded, aged, or otherwise disabled patient might impose too large a burden on society, or family, or on the patient himself. The doctor might see his role as an agent of social and economic good as well as the patient's "good." In such cases he might refuse to satisfy the particular request for CPR, or perform it ineffectively to mollify the staff, or family, or to avoid legal liabilities. Even less commendably, the physician might allow his own estimate of a patient's moral or social worth to influence his decision to withhold CPR from patients he considers social "misfits" - sociopaths, chronic alcoholics, vagrants, criminals, or drug addicts. In these instances the good of the patient is usurped by the physician's concept of good.

IV. SOCIAL GOOD

We have concentrated on the meanings of the patient's good because this is the *raison d'être* of all clinical medicine. But as the cumulative effects of individual medical decisions alter the world's demography, and high technology consumes an ever larger percentage of its resources, the other good things a society seeks come more and more into conflict with the good of individual patients. Already, the economic and social costs of prolonging the lives of many aged and disabled patients are intruding themselves into clinical decisions. Medicine's more vocal critics often take it to task for over-emphasizing the good of the individual patient and neglecting the good of society. As a result, the principles of social and distributive justice are increasingly invoked in clinical decisions.

With respect to CPR, some questions of justice can be legitimately raised: Is the patient really free to choose CPR simply to achieve something he deems in his best interest, i.e., arranging his affairs, seeing a grandchild graduate from medical school? Are not the costs he incurs a burden on others? Recall that cardiopulmonary arrest outside the hospital is generally considered an automatic indication for CPR since neither rational decisions nor proper diagnosis and prognosis are possible under such urgent conditions. Is this policy tenable if the probability of eventual survival or return of socially active life is small and the costs to society high? Is the surrogate of a severely impaired infant or neonate free to choose CPR if that infant's future life is to require the use of extensive resources, and expenditures which the patient cannot afford and the whole of society must bear? What if these expenditures are likely to continue long after the proxy is able to help in defraying them?

These questions bring the patient's interests and even his agency squarely into conflict with the good of others. Some advocate a form of social paternalism to protect society against the economic and social consequences of CPR in seriously ill patients. The possibility of a public policy in the U.S. that would deny CPR to certain categories of patients is not beyond contemplation.

As a result, physicians are being pressured to assume the role of monitors of the social and economic good of society. When patient good and social good are thus intermingled they may well come into conflict. How consistent is a concern for social good with the traditional ethical commitment of physicians to do everything to advance the interests and medical good of their patients?

The foregoing questions illustrate the complexity of the concept of the good of the patient. They underscore the need not only for more explicit analysis of the points of conflict but also for their resolution. That resolution in its turn requires establishment of some hierarchical order among the various notions of *good* if decisions are to be made on any principled basis.

V. MAKING MORAL CHOICES - THE HIERARCHICAL ORDER OF PATIENT GOOD

Making clinical decisions with moral implications necessitates some ranking of the four senses of patient good because they can come into conflict. I would now like to examine more closely the hierarchical relationship which should obtain among these several levels of patient good when they are in conflict. Conflicts between good things can only be resolved according to some rational organizing principle.

Some of us clearly organize our lives around one good that is the highest good in our lives, traditionally the *summum bonum*. For those who reject that idea, I would refer simply to the "good of last resort" - that good to which we tend to return whenever we are forced to make choices between competing goods, the one good we tend to place above others. For the religious person the highest good is accommodation to the will or law of the Creator. For the non-religious person it may be seeking the greatest pleasure, the least harm, the greatest utility, enlightened self-interest, the good of the least advantaged person in a society, the absolute autonomy of patients to choose or the survival of the species. Without arguing at this point about what that ultimate good should be, we need only accept that there is *de facto* such a good for all who attempt to make rational choices. Even pluralistic intuitionists must base their intuitions on a final good, in this sense.

Clearly, the ultimate good, or the good of last resort, will take precedence over the other forms of patient good. Strong paternalism with respect to a patient's choice of ultimate good is morally offensive. The ultimate good is the starting point of a person's moral reasoning, his first act of intellectual faith so to speak. If he or she is competent it must be respected over medical good, and the physician's, society's, the family's, or the law's construal of ultimate good. The patient may abandon or subjugate his conception of ultimate good to his more immediate personal interests, but others may not do so.

There has never been, nor is there likely ever to be, universal agreement on the ultimate good. Societies that wish to be homogeneous in their choice of ultimate good usually do so by some form of coercion. In democratic societies it is a civic right of competent persons to choose their own belief systems. To pursue a moral life we are under compulsion to act with fidelity to some ultimate source or concept of good though our choices of that source or concept may vary widely.

When conflicts occur in decisions involving human life - as for example with no-code orders, discontinuing life support measures, artificial insemination, abortion, etc. they often involve disagreements about the ultimate good and are therefore reconcilable with great difficulty if at all. Under such circumstances the patient-physician relationship should be respectfully and courteously discontinued since neither physician nor patient can morally compromise his belief system - particularly when the issue involves ultimate good.

The good next in order of priority is the good of the patient as a human person, his freedom to make his own decisions. To place even medical good or the quality of life ahead of freedom to choose is to rob the patient of his humanity. The physician instead has the obligation to enhance the patient's competence in every way - treating pathophysiological disturbances of brain function, freely providing accurate information needed to make choices, and refraining from coercion or deception even to overcome resistance to needed treatments.

The once-competent patient who at the time of decision is comatose, psychotic or otherwise unable to reason and choose does not lose his claim as a human being to have his interests respected. We turn therefore to surrogates, who can act on behalf of those interests. Presumably the surrogate knows the patient better than the physician, or is closer to his cultural or ethical value system. His choices are more likely to approximate what the patient would have wanted were he able to express his choice. If the surrogate or proxy is competent and is clearly acting in the interests of the patient, his or her choices take precedence over the physician's. In the difficult situation in which there is doubt about the capability or good intention of the surrogate, then the physician must attend to the good of the patient. This may require resort to the courts if the physician thinks the surrogate's decision is not in the patient's interests. Under these circumstances, at least the patient's legal rights can be protected, and the intentions of both the surrogate and the physician examined a little more objectively, recognizing always that a legal decision may not be necessarily a morally correct one.

In the case of the never-competent - infants and the retarded - we have no way of knowing what the patient would have chosen, nor do the surrogates, even when they are the parents. Here there is no way to determine what is or would be the patient's perception of his own good, or of his ultimate good. Neither will the patient ever be free to choose. Under these circumstances three of the four components of patient good, as we have been discussing them, cannot be ascertained. Biomedical or techno-medical good therefore assumes a dominant role. It is probably a safer guide than presumptions about what an infant would consider in his interest as his life evolves in the presence of a severe handicap I would, under these circumstances, tend to Paul Ramsey's position, i.e., if some clinical benefit, easing of pain, discomfort, or physiological improvement can be obtained for the patient, then treatment probably should be undertaken ([25], pp. 189-227). Obviously we cannot know at the time of decision whether in the long run the patient will thank us or not. But at least, the decision will be taken on the basis of what can be known, and not on what can only be presumed with the greatest uncertainty about another person's future preferences.

With the never-competent patient, it is still the good of the patient that ordinarily remains our concern - not that of the parents, or guardians, or society in general. The possibilities for disagreement on what is "best" in such cases are evident from the wide range of arguments in the current literature. For some, spiritual good may accrue to the family and society from the care of a deformed, defective or disabled infant. For a utilitarian the social costs of a life of total dependency may be totally unjustifiable. Obviously the case of the never-competent will occasion the greatest difficulty in applying any hierarchical order to patient good. Yet some attempt must be made, otherwise decisions will be unpredictable, intuitive and fortuitous.

The choice of competent surrogates is limited if it violates the conscience of the medical attendant or contravenes any of the senses of patient good we have outlined. The physician or nurse must remain faithful to his or her conscience. His or her humanity is as precious as that of the patient. Coercion by court order, law, or other means of the physician's decision is as indefensible as coercion of the patient's decision.

All the moral obligations that derive from this schema of patient good are obviously not spelled out here. All health professionals, health care institutions, ministers, families and friends - all who participate in decisions that affect those who are ill - must take into account the many dimensions of "patient good." In the best decisions, the four senses I have outlined should be closely congruent with each other. When they are not, the differences should be spelled out as clearly as possible, and negotiated with honesty, and sensitivity. This is an instance in which the quality of the personal relationships and the character traits of the interacting parties will often be more important than rules or procedures.

The hierarchical order I have suggested has implications for the way the common principles of medical ethics are applied in particular instances. This is pertinent for each of the several varieties of paternalism ([8], pp. 16-22. The physician who is a strong paternalist, for example, might place techno-medical good above other senses of patient good. For him the patient's perceptions of his own immediate or ultimate good and his freedom, as a person, to make his own choices, are not primary concerns. For some strong paternalists techno-medical good justifies withholding or manipulating information, or breaking promises. Even the use of deceit or force might be rationalized to assure compliance with what is medically indicated.

Conversely, the libertarian typically places personal autonomy at the head of the list of what is good for the patient. A Rawlsian contractarian would opt for the good of the least advantaged member of society, though a Hobbesian would not. For the utilitarian social good would seem most defensible. Some sociobiologists deem the preservation of the gene pool the highest good. Some economists might choose just distribution of scarce resources and others the unimpeded operation of the free market.

The way the principles of beneficence, non-maleficence, justice, truth-telling and promise-keeping are applied is linked to the interpretation we put on the good of the patient. The hierarchical order we choose for arranging the four senses of patient good will differ with the ethical theory we espouse because every theory is grounded in some concept of the good. That order may vary even with different fields of medicine. For example, in preventive medicine, and public health as contrasted with curative medicine, certain strictures must be placed on the patient's expression of his own interests and even his freedom to refuse certain interventions like immunization [2].

VI. PATIENT GOOD AND ARISTOTLE'S NOTION OF GOOD

These problems are further exemplified when the question of patient good is set in the context of contemporary views of the good and contrasted with Aristotle's classical account.[1] In a recent study Henry Veatch, in part following Leo Strauss, contrasts the teleology of the modern view - of contractarians and consequentialists - with that of Aristotle [29]. The "moderns" interpret morality in terms of rights, particularly the right to define good for oneself limited only by the rights of others to do likewise. Rights then precede duty and morality enters when the pursuit of one's own ends and interests conflicts with the same pursuit by others. On the "modern" view we cannot know what is good for the patient without knowing his desires. The patient's choice is a good simply because he desires it. To do good for the patient we should do the good he desires. This concept of good is most consistent with a libertarian or permissive, anti-paternalist, stance that permits the patient to choose a course even when it may seem wrong to a reasonable observer so long as the consequences are not harmful to others.

On the classical-medieval view, the good is objective and intrinsic to things that are good. These things ought to be done because they perfect our humanity and are most fitting for humans as humans. Good therefore is to be determined without reference to whether the patient wants it. Indeed he *ought* to want it and has a duty to want it. Duty precedes rights. Indeed rights exist because we must be unimpeded to do the good we ought to do. Patients ought not to be free to choose an evil course of action. For their good, their wishes can be overriden.

Veatch's interpretation of Aristotelian teleology may slide somewhat too easily over the controversies about precisely what Aristotle meant by the good. Nicholas White, for example, holds that Aristotle used good in the *Ethics* in at least three senses: (1) that which is good in itself and desired for itself and for which all other things are sought, (2) that which is good for a human being, or (3) that which is the aim or desire of an individual [30]. White attempts valiantly to reconcile these views in terms of *the good* but his efforts are not entirely convincing. It seems far more likely that Aristotle too had to confront the fact that humans do in fact see the good in at least these four ways and that they are probably not completely reconcilable with each other.

W. F. R. Hardie on the other hand suggests that Aristotle's doctrine of the good may be taken in two ways: the first he calls "inclusive," i.e. encompassing a wide variety of aims consistent with the kind of life one wants to live [14]. Man is distinguished from other creatures as the responsible planner of his own life. His choices therefore are apt to be quite individual and peculiar to each person's plan of life. Hardie's second idea is of good as the dominant end. Here good is the overall plan that most fully makes a man a man, i.e., the attainment of theoretical knowledge. Hardie says "Aristotle's doctrine of the final good is a doctrine about what is proper to a man. The power to reflect on his own abilities and desires and to conceive and choose for himself a satisfactory way of life'' ([13], p. 321). This is reminiscent of Ralph Barton Perry's general theory of value which distinguishes man from animal bv his ability to plan ahead in accord with his interests [24].

Let us return, briefly to the three senses of 'good' in Aristotle listed by White: (1) good *simpliciter,* that is, ultimate good for which all other things are chosen; (2) good for someone, i.e., a particular person; and (3) good for a human being. I would like to suggest a correspondence between these three senses of 'good' and the four senses of *patient good* I have offered.

Aristotle's good *simpliciter* is, of course, *the good* or ultimate good. It is always in the background in clinical decisions. Even those who deny the usefulness of such a concept nonetheless tend to have a ''good of last resort,'' so to speak, on which they found their sense of what is morally defensible.

It is the second and third senses of good, as used by Aristotle, i.e., the good of a particular person and the good of a human being, that correspond more closely with the several senses of patient good I have discussed.

The ''patient's interests,'' his idea of what is good for him and fits his life plan, corresponds rather well with Aristotle's notion of the good as something at which people aim, the good of particular persons. Medical good, is then, the instrumental good that enables the patient to achieve his aims given the exigencies of illness.

"Freedom to choose" corresponds to Aristotle's third sense of the good as what is proper to a human being - someone endowed with reason ''... for man therefore the life according to reason is best and pleasantest since reason more than anything else is man'' ([4], 1178a 6-8). This passage has been interpreted as pointing to the life of contemplation, but as Hardie argues, it also points to the good of choosing, planning one's own life, and defining one's selfhood. Without freedom to choose it is impossible to use one's reason in an effective way.

Preeminent realist that he was, Aristotle's struggle with the conception of the good probably reflects his sensitivity to the actual difficulties of any simple definition of such a compound term. The realities of clinical decisions and the trifold notion of patient good argue somewhat against those who attempt to conflate Aristotle's three conceptions into one.

Medical good, the goal of medical interest, namely the health of the body, was for Aristotle a subsidiary good. It was a necessary condition and a means for the pursuit of the good life and the ultimate good but not itself ultimate [21]. Thus in the *Rhetoric* he says, "The excellence of the body is health; that is a condition which allows us while keeping free from disease to have use of our bodies" ([3], 136lb3). But he did not think health should override other human concerns, ". . . for many people are healthy as we are told Herodotus was, and these no one can congratulate on their health for they have to abstain from everything or nearly everything that men do" ([3], 136lb4). Aristotle's assignment of health in the hierarchy of good is not very far from the position we have given medical good in our analysis.

VII. RECAPITULATION

Everyone who participates in a clinical decision justifies his or her actions as being in behalf of the patient's good. I have tried in this essay to clarify that all-inclusive notion by analyzing it into four components, identifying conflicts among these components as interpreted by the patient, the physician, the family and society, and suggesting how these conflicts might be resolved in making clinical decisions. I have confined myself to the patient's good as perceived by the patient and have avoided the more profound issues of the good of the patient metaphysically. I have used the example of decisions not to resuscitate as a paradigm case for the issues involving the patient's good though the analysis is applicable to most clinical decisions involving moral choice. Finally I have suggested that a proper analysis of the patient's good will make for an ethically sounder physician-patient relationship, and a clearer interpretation of the usual principles of medical ethics.[2]

Georgetown University
Washington, D.C.

NOTES

[1] See [4], especially Book I; Book X, Chapters 2, 7-8; and Book XII, Chapters 6-7, 9-10

[2] The author owes special thanks to the editors, Loretta Kopelman and John Moskop, for their careful editing and helpful suggestions which are incorporated into the text.

BIBLIOGRAPHY

[1] Annas, G. J.:1982, 'CPR: The Beat Goes On', *Hastings Center Report* **12:** 4 (August), 24-25.

[2] Annas, G. J.: 1982, 'CPR: When The Beat Should Stop', *Hastings Center Report* **12:** 5 (October), 30-31.

[3] Aristotle: *Rhetoric.*

[4] Aristotle: *Nichomachean Ethics.*

[5] Brandt, R. B.: 1979, *A Theory of the Good and the Right,* Clarendon Press, Oxford.

[6] Brock, D. W.: 1983, 'Paternalism and Promoting the Good', in R. Sartorius (ed.) *Paternalism,* University of Minnesota Press, Minneapolis, pp. 237-260.

[7] Carson, R. A. and Siegler, M.: 1982, 'Does "Doing Everything" Include CPR', Case Study, *Hastings Center Report* **12:** 5 (October), 27-29.

[8] Childress, James F.: 1982, *Who Should Decide? Paternalism in Health Care,* Oxford University Press, New York.

[9] Clinical Care Committee of the Massachusetts General Hospital: 1976, 'Optimum Care for Hopelessly Ill Patients', *New England Journal of Medicine* **295:** 7, 362-369.

[10] Erde, E. L.: 1978, 'The Place of the Good in the Science and Art of Medicine', *Man and Medicine* **3:** 2, 89-100.

[11] Frankena, W. K.: 1963, *Ethics,* Prentice Hall, Englewood Cliffs, N.J.

[12] Gruzalski, B.: 1982, 'When to Keep Patients Alive Against Their Will', in B. Gruzalski and C. N. Ballinger (eds.), *Value Conflicts in Health Care Delivery,* Ballinger Publ. Co., Cambridge, Mass., pp. 171-191.

[13] Hardie, W. F. R.: 1967, 'The Final Good in Aristotle's Ethics', in J. M. E. Moravcsik (ed.), *Aristotle: A Collection of Critical Essays,* Anchor, Garden City, N.Y., pp. 296-322.

[14] *In re Dinnerstein,* 380 N.E. 2d 134 (Mass. App. 1978).

[15] Jackson, D. L. and Youngner, S.: 1973, 'Patient Autonomy and Death with Dignity', *New England Journal of Medicine* **301:** 8, 404-408.

[16] Lavelle, L.: 1972, *The Problem of Evil in Contemporary European Ethics,* J. J. Kochelmans (ed.), Anchor Books, Garden City, N. Y.

[17] Levine, R. J.: 1981, 'Do Not Resuscitate Decisions and their Implications', in C. Wong and J. P. Swazey (eds.), *Dilemmas of Dying: Policies and Procedures for Decisions not to Treat,* G. K. Hall, Boston, pp. 23-41.

[18] Miles, S. H., Cranford, R. and Schultz, A. L.:1982, 'The Do-Not-Resuscitate Order in a Teaching Hospital', *Annals of Internal Medicine* **96:** 5, 660-664.

[19] Olson, R. G.: 1972 'The Good', in P. Edwards (ed.), *The Encyclopedia of Philosophy,* Free Press, Macmillan Publishing Co., Vol. 3., pp. 367-370.

[20] Owens, J.: 1977, 'Aristotelian Ethics and Medicine', in S. F. Spicker and H. T. Engelhardt, Jr. (eds.), *Philosophical Medical Ethics: Its Nature and Significance,* D. Reidel Publishing Co., Dordrecht, Holland, pp. 127-142.

[21] Pellegrino E. D.: 1983, 'Autonomy and Coercion in Disease Prevention and Health Promotion', *Theoretical Medicine* **5:** 1, 83-91.

[22] Pellegrino, E. D.: 1979, 'Towards a Reconstruction of Medical Morality: The Primacy of the Act of Profession and the Fact of Illness', *Journal of Medicine and Philosophy* **4:** 1. 32-56.

[23] Pellegrino, E. D.: 1985, 'The Virtuous Physician and the Ethics of Medicine', in E. E. Shelp (ed.), *Virtue and Medicine,* D. Reidel Publ. Co., Dordrecht, Holland.
[24] Perry, R. B.: 1926, *General Theory of Value,* Longmans, Green, New York.
[25] Ramsey, P.: 1978, *Ethics at the Edges of Life,* Yale University Press, New Haven, pp. 189-227.
[26] Ross, W. D.: 1930, *The Right and the Good,* Oxford University Press, London.
[27] Spencer, S. S.: 1979. 'Code or No Code: A Nonlegal Opinion', *New England Journal of Medicine* **300**: 3, 138-140.
[28] U. S. President's Commission on Ethical Problems in Medicine and Biomedical and Behavioral Research: 1983, *Deciding to Forego Life-Sustaining Treatment,* Washington, D.C.
[29] Veatch, H. B.: 1981. 'Telos and Teleology in Aristotelian Ethics', in D. O'Meara (ed.), *Studies in Aristotle,* Catholic University of America Press, Washington, D.C., pp 279-296.
[30] White, N. P.: 1981, 'Goodness and Human Aims in Aristotle's Ethics', in D. O'Meara (ed.), *Studies in Aristotle,* Catholic University of America Press, Washington, D.C., pp. 225-246.
[31] Wong, D. C. and Swazey, J. P.: 1981, *Dilemmas of Dying: Policies and Procedures for Decisions not to Treat,* G. K. Hall, Boston.

SALLY A. GADOW

WHAT GOOD IS ANOTHER PAPER ON THE GOOD? NO CODES AND DR. PELLEGRINO

It is startling – and refreshing beyond measure – to find in Edmund Pellegrino's paper a discussion of codes and goods in which (a) there is no dependence upon the ordinary/extraordinary distinction; (b) the author unabashedly places biomedical good at the bottom of the hierarchy of goods; and (c) the by now wearisome debate over autonomy and paternalism is dispatched with the assertion that both camps are fueled by the same commitment, the good of the patient. However vehement the debate may become over meanings of "the good of the patient," it is a welcome advance not only to recognize that a concern for patient good is behind both positions, but to engage in a patient analysis of this "most ancient" principle and the "vexing ethical dilemmas" it engenders ([3], p. 117) – knowing well that few topics are greeted with greater impatience than those that are ancient but still vexing.

Pellegrino frames his analysis of patient good in the context of critical care, specifically in terms of decisions about cardiopulmonary resuscitation (CPR). This is a useful context for analysis, since for him the decision not to resuscitate is "a paradigm case" whose "analysis is applicable to most clinical decisions involving moral choice" ([3], p. 136), i.e., there is nothing morally distinctive about decisions to code or not to code patients. These decisions simply display, perhaps in the starkest way, issues immanent in all clinical decisions-the most important of which, he contends, is the issue of how to determine the patient's good. That determination is more critical in CPR than in other decisions because it can lead to extreme and irreversible consequences: to death or to an indefinite decerebrate condition. The stakes are high, and he rightly signals the determination of patient good as the central issue to be addressed.

Pivotal in Pellegrino's analysis is the distinction among three senses of patient good (leaving aside his category of ultimate good): medical good, personal interests, and individual choice. With these distinctions in hand, it becomes impossible innocently to conflate, equivocate, or otherwise fail to distinguish meanings of good. An example of such failure is the justification offered by Siegler of an emergency decision not

J. C. Moskop and L. Kopelman (eds.), Ethics and Critical Care Medicine, 139-145.

to resuscitate a previously alert patient who had given no indication of readiness to discontinue treatment and had not been consulted about his wishes regarding resuscitation: "*perhaps* the physician acted rightly in not raising this matter with Mr. Walker (even if Mr. Walker were competent). If the physician knew . . . that CPR offered Mr. Walker no possible benefit, the patient's preferences would not have been germane" ([1], p. 29). Siegler's "no possible benefit," while purportedly referring to (and thus conflating) any and all senses of benefit, can in fact refer only to medical benefit, since neither of the other two senses of patient good – personal interests or individual choice – could have been ascertained *without* raising the matter with Mr. Walker.

Pellegrino, it is fair to assume, would not offer Siegler's justification for withholding CPR, since he maintains that medical benefit is the least significant of the three senses of patient good. He admits that there may be discrepancy among the different senses of benefit. Although CPR may offer little or no medical good, nonetheless it may be desired by a patient who values life to the last moment. In that case, without the third sense of patient good (the value of individual choice) conflict between the medical uselessness of CPR and the patient's interest in life could produce a stalemate broken only by the physician either agreeing with Pellegrino that the medical value of the procedure is the least relevant of the possible goods or agreeing with Siegler that it is the most relevant. Introduction of the third sense of patient good, the dignity of self-determination, clearly weights the balance in favor of ordering a code if the patient wishes it. Notice that the good affirmed here is not the resuscitation itself. Pellegrino recognizes that something is not made a good just by being chosen. The good that the decision affirms is the good of liberty or choice itself.

But here a problem arises with the three senses of good. Why do we need all three? Why do we need the concept of medical good if "in medical decisions, so long as the patient is competent, we must allow him to make his own choices" ([3], p. 125)? And why distinguish the latter two senses, since they are two facets of the same good, autonomy? The claim that personal interests can define the good for an individual presupposes that personal choice regarding one's interests is also good. Is Pellegrino then advocating any good beyond that of freedom, which is the prerequisite for patients' subjective determination of personal good and beside which objective (medical) senses of good barely count? Is he simply offering a reformulation of the autonomy position flavored with paternalism: if freedom of choice is the good to which the three senses of good can be reduced, and if we agree with

Pellegrino that the good of the patient "is the ultimate court of appeal for the morality of medical acts" ([3], p. 117), is there an obligation to see that patients act autonomously because so acting is *for their own good?*

If there is such an obligation, it may be agonizingly hard to fulfill. In their discussion of CPR decisions, Jackson and Youngner [2] observe that it can be exceedingly difficult to ascertain whether the good of patient automony has been attained in a given case. Consequently, if faced with the choice of trying to bring about an ill-defined, elusive, often unattainable good for a patient or a good that is easier to provide (patient equanimity or trust, for example), does one not do more good for the patient by attempting the latter and succeeding than by attempting the former and failing? That is the apparent implication of the concern not to harm patients by promoting their autonomy: "it may be inappropriate to introduce the subject of withholding cardiopulmonary resuscitation efforts to certain competent patients when, in the physician's judgement, the patient will probably be unable to cope with them psychologically;" in those cases, presumably, CPR will be provided, since the patient has not expressed a preference for the withholding of resuscitation ([4], p. 365). There is equal concern, however, when patients do cope with the subject of CPR decisions and articulate a clear preference for or against being resuscitated: "Physicians who are uncomfortable or inexperienced in dealing with the complex psychosocial issues facing critically ill patients may ignore an important aspect of their professional responsibility by taking a patient's or a family's statement at face value" ([2], p. 407). Finally, to compound the problem of autonomy as the reigning good in CPR decisions, Jackson and Youngner point to the possibility that at times a physician accedes to a patient's decision to die because he or she believes the patient better off dead: "one must be cautious not to act precipitously on the side of the patient's ambivalence with which one agrees, while piously claiming to be following the principle of patient autonomy" ([2], p. 407).

When Pellegrino's two senses of patient good having to do with autonomy are difficult to attain, and if one is committed to doing *some* good for the patient, then what is left? Pellegrino's notion of medical good, perhaps too hastily discounted earlier, must be examined. Pellegrino attempts to rid the notion of any evaluative, quality-of-life overtones, stripping it to the barest objective judgement regarding the effects of medical intervention on the natural history of the disease being treated or the probability of ameliorating the patient's disability or discomfort ([3], pp. 121-122). Medical good, in short, is an instrumental good for the *benefit* or *improvement* of the patient's situation, by either curative or palliative intervention. Without hope of positive change regarding either the pathological process or the individual's suffering, intervention is not justified, medical benefit is unattainable.

Using that criterion of medical good, CPR is indicated only when it carries the possibility of improving the patient's condition. Here it becomes essential to determine exactly which condition CPR is to correct, (1) the immediate crisis itself (cessation of vital functions) or (2) the preexisting pathology or suffering. Clearly, CPR affects (2) only indirectly by reversing (1). The question thus is: Should the expected effect upon (2) be the primary or the secondary measure of medical good? Pellegrino seems to view it as the primary measure: CPR offers no medical benefit in the case in which a patient can be returned only to a hopeless (unimprovable) condition. This evidently is Siegler's view regarding CPR for Mr. Walker.

This view of the medical benefit of CPR has important consequences. It means that measures conducive of medical good include not only those that directly effect an improvement, but also other, indirect measures – most notably, diagnostic procedures – that make possible the subsequent interventions which will directly affect the patient's condition. If there are no direct measures available, as in a hopeless condition, none of the indirect measures are justified, since their only benefit lies in improving the patient's situation and that has been judged impossible to to.

One way of clarifying Pellegrino's view may be to consider how it relies upon an understanding of what is hopeless or what is useful. The failure to provide CPR on the grounds that it will be of no medical good could be like refusing to administer lifesaving aid to a blind person who has walked in front of a car on the grounds that it would not affect her blindness. Now the problem with Pellegrino's concept of medical good becomes evident. It has not in fact been freed of quality of life considerations, but contains an implicit understanding of what is meant by "hopeless." This is crucial, moreover, since hopeless conditions are those for which both direct and indirect (including resuscitative) measures can be justifiably withheld. Blindness, though incurable, is not a hopeless condition since there are measures available to provide the sightless person with a life of acceptable quality, i.e., a life without unmanageable impairment. But clearly, when dealing with a competent person, the judgment that a situation is bearable is a subjective one that individuals can make regarding *only their own lives.* Even in a typically "hopeless" condition such as a terminal illness, the time remaining may be invaluable to the individual despite significant suffering. A situation that is clinically hopeless - without possibility of reversing either the pathology or the suffering - may not be personally hopeless. Pellegrino himself would insist upon this distinction between the external judgment of a situation and the subjective experience and evaluation of it. For his distinction between medical good and autonomy, however, the difficulty is clear: the judgment about medical benefit involves an irreducible element of subjective evaluation by the patient and thus cannot be separated from the exercise of patient self-determination. Withholding CPR when it will have no medical benefit, i.e., when it can return the patient only to a clinically hopeless condition, obviously avoids the often formidable problems of ascertaining a patient's wishes, but more significantly, it negates in principle the essentially personal determination involved in the designation of a situation as hopeless and without value. The concept of medical good fails after all to provide an alternative to patient autonomy as a means of deciding whether to intervene, with either direct or indirect measures.

I would like to propose another way in which the good of CPR can be understood, viz., rescue. CPR as rescue is direct intervention to alleviate, not the underlying or prior condition, but only the presenting crisis, the cessation of vital functions.

I have applauded Pellegrino's avoidance of the ordinary-extraordinary distinction in a discussion of CPR, but the distinction turns out to be important in an unexpected way, indeed a reversal of the way the distinction usually is employed. As rescue, CPR is not like other, "ordinary" procedures that are medically justified only if they are of at least indirect therapeutic value. Resuscitation is "extraordinary," not because it is exotic or out of the ordinary, but precisely because it is - as rescue - the most ordinary of human responses, ordinary in the sense of immediate, automatic, unquestioned, requiring no justification beyond the knowledge or presumption of the patient's wishes. Its purpose is not therapeutic in the usual meaning of the term; its purpose is to restore an individual to a condition in which the therapeutic question *then* can be asked about whether or not treatment is medically indicated and personally desired. Its only "benefit" is that, if successful, it restores a situation in which other goods can be offered, goods in any of Pellegrino's senses. What goods, then, are served by CPR as rescue? The personal good attained depends upon the patient's desire for continued life. If that desire cannot be clearly or quickly enough established to count as a good, there is no therapeutic good upon which to fall back in seeking a justification for the procedure. Consequently, on this view CPR is a measure that may not be withheld when it offers possibility of improving the patient's pre-existing condition.[1]

The dazzling technology of resuscitation fosters the assumption, of course, that CPR is a form of treatment. Other confusions follow, including the notion of "full codes" and the more insidious notion that dignity and technology are opposed in terminal care. The technology notwithstanding, resuscitation is extraordinary only in its utter ordinariness, ordinary in the same way that the provision of food for someone starving is ordinary human care rather than a medical procedure requiring therapeutic benefit for its justification.

There seem finally to be only three alternatives: (1) Decisions about orders to resuscitate can be made without involving patients in the deliberation, since patients' wishes may become hopelessly clouded or skewed – for example, by their discovery that others are willing to let them choose death. (2) Patients can be involved in the deliberation, with the realization on the physician's part that the promotion of patient autonomy becomes an abuse when it serves as a camouflage for abandoning the patient. (3) CPR, like other forms of rescue, can be considered morally obligatory for anyone capable of performing it, except in those cases in which there is indisputable evidence that competent persons do not wish to be rescued.

Perhaps Pellegrino has not after all avoided the paternalism issue, but instead has adroitly shown that any discussion of The Good must be a discussion of the choosing among goods, and that the goods of resuscitation may be related to neither medical benefit nor patient autonomy but the common trust that rescue is, for the physician no less than for the friend and even the stranger, one of the most taken-for-granted forms of human response.

The University of Texas Medical Branch
Galveston, Texas

NOTE

[1] This is not an argument of life at all costs, for it may be that there are no goods that the patient accepts as consolation for the suffering of remaining alive. The resuscitation of a person known to hold such a view obligates the rescuer to provide the person with a good, beyond the restoration of life per se, that compensates the individual for the forced resumption of suffering.

BIBLIOGRAPHY

[1] Carson, R. A. and Siegler, M.: 1982, 'Does "Doing Everything" Include CPR?' *Hastings Center Report* **12:** 5 (October), 27-29.
[2] Jackson, D. L. and Youngner, S.: 1979, 'Patient Autonomy and "Death with Dignity": Some Clinical Caveats', *New England Journal of Medicine* **301,** 404-408.
[3] Pellegrino, E.: 1985, 'Moral Choice, the Good of the Patient, and the Patient's Good', in this volume, pp. 117-138.
[4] Rabkin, N. T., Gillerman, G., and Rice, N. R.: 1976, 'Orders Not to Resuscitate', *New England Journal of Medicine* **295,** 364-367.

JOHN C. MOSKOP

ALLOCATING RESOURCES WITHIN HEALTH CARE: CRITICAL CARE VS. PREVENTION

Once we as physicians start making decisions regarding who is or is not worthy or deserving of our best efforts, we revert to black and white magic. Physicians will then see again fear in the eyes of their patients - not fear of pain or fear of death but fear of the *physician* .... Because of the nature and costs of our practice, we intensivists are the first to feel such pressure to bend our principles. We must also be the first to resist ([14], p. 169).

This is the strong response of Dr. Louis Del Guercio to the prospect of cutbacks in federal support for critical care, taken from his 1977 presidental address to the Society For Critical Care Medicine. Dr. Del Guercio's remarks illustrate one side of a continuing controversy about how much society ought to invest in critical care medicine, the care of patients with severe, life-threatening illness or injury. This paper will be devoted to an examination of that controversy. First, however, I would like to place the controversy within the broader context of concern about the cost of health care.

In the United States, the establishment in 1965 of the federal Medicare and Medicaid programs to subsidize care for the elderly and poor, may, for a time, have created the illusion that all real needs for health care in this country could and would be met. This, I suspect, was never more than an illusion; there have always been barriers to care for some groups of people, including those too far away from facilities, those who fell into the gaps in our patchwork system of health insurance, and those who did not know that their problems were amenable to treatment or did not know how to obtain medical attention. The constant and rapid rise in health care costs over the past 15 years, however, has finally forced recognition of the extreme difficulty, if not the impossibility, of satisfying all needs for health care. The Reagan administration has responded with across-the-board cutbacks in federal funding for health care programs and most recently with the prospective reimbursement system for Medicare based on diagnosis-related groups (DRGs).

Many disagree with these current approaches to containing health care costs, but almost no one currently writing on this topic denies the need to limit and control in some way the amount of resources devoted to health

J. C. Moskop and L. Kopelman (eds.), Ethics and Critical Care Medicine, 147-161.

care. How to change our current system of health care in order to control its costs is, however, an extremely complex issue. It obviously involves many questions about the effectiveness and efficiency of particular kinds of care. (For example, do most tonsillectomies have a positive impact on health? Are medical or surgical approaches a more cost-effective way to treat atherosclerosis?) Other questions of ethics and values, however, have also played a significant role in determining health care delivery policies at several levels. At the most *general* level, policy-makers seek to compare the values obtainable through all kinds of health care with the values of projects and programs in other areas of national endeavor, e.g., defense, education, welfare, conservation and recreation. At an *intermediate* level, different areas within health care have been compared in terms of potential benefits, risks and costs in order to allocate resources among the various kinds of health care most rationally. At a more *specific* level, different procedures have been proposed for determining which individuals should receive a particular scarce medical resource when there is not enough for all who need it.

Thus, issues of cost control and allocation in health care include a very broad range of questions. In this paper, I will focus on only one part of that range, the intermediate level of allocating resources among different areas within health care, and then only on a few of those areas. That is, I will discuss arguments for and against giving special priority within health care to the provision of critical care. I will also take a briefer look at arguments for increased support of preventive medicine. My examination will focus on conceptual and moral issues. Though I will not be able to state all of the arguments in full detail, I will try to capture their strongest features. Nevertheless, I will argue that each of the arguments has a serious weakness. I will, therefore, conclude that thus far, neither the proponents of critical care nor those of preventive medicine have made their case convincingly. Though my conclusion is a negative one, it will, I hope, clear the way for new contributions to the debate.

I. IN SUPPORT OF CRITICAL CARE

Del Guercio and a number of other writers seek to defend the established priority on critical care in Western health care systems [7, 14, 18, 22]. Some have identified features of critical care which are claimed to justify the priority of critical care *even if* preventive approaches would make a greater overall contribution to health in the long run. Others have pointed out that certain kinds of critical care, at least, are very effective and have a tolerable cost. I will examine, first, those arguments which seek to justify the priority of critical care even if other approaches would make a greater overall contribution to health in the long run. These arguments are based on (1) the symbolic value of critical care; (2) the physician's commitment to individual patient welfare; and (3) the urgency of critical care needs.

The *first* argument in support of critical care is based on its *symbolic value*. A number of writers [5, 7, 18, 20] have argued that society benefits from the symbolic significance of providing resources for life-saving care at a level sufficient to make such care available to all who need it. According to Calabresi and Bobbitt, "since many other values are constantly being eroded by decisions which, in fact, place a low value on human life, substantial benefits accrue from any demonstration by society of its devotion to life's pricelessness" ([3], p. 135). The decision to support critical care is said to be beneficial even if investment of the same resources in simple preventive measures would, in the long run, save a greater number of lives. This is so, it is argued, because the lives lost due to lack of prevention are not individually identifiable. That is, each individual death can be attributed directly to some specific cause, but only indirectly and more conjecturally to the absence of a preventive measure. Thus, we can attribute a man's death directly to a heart attack, but only indirectly and more conjecturally to the absence of a hypertension screening program. Individual deaths which might have been avoided by means of preventive measures may, then, more readily appear to be unavoidable accidents or "acts of God" than the results of calculated social choices.

Proponents of the symbolic value argument like to compare critical care to other expensive lifesaving efforts, such as the rescue of trapped miners or of persons lost at sea. What actually is gained by making critical care a symbol for commitment to the value of life, however, is not usually clearly stated. In fact, what these advantages might be and how they would be realized are far from obvious. It might, for example, be claimed that a widespread belief, triggered by the provision of critical care, that society places a high value on life will cause individuals also to place a high value on life and to reflect this in their actions. No evidence is cited to support this, however, and others have argued that public provision for personal needs may actually decrease the motivation of individuals to undertake altruistic or charitable action ([24], p. 8;

[23]). Defenders of the symbolic value of critical care might also argue that the availability of critical care inspires feelings of approbation toward society and of personal security. We may feel gratitude and loyalty toward our society for its apparent commitment to the value of life, and we may also feel a greater sense of protection from death because of the availability of critical care. Such feelings, however, if they do in fact exist, would be based on a kind of deception. By refusing to fund preventive measures which could save many lives, society *does* put a price on life, though the fact that more visible services like critical care are provided may prevent individuals from recognizing this. Likewise, though it may help to prevent anxiety, the sense of personal security engendered by a reliance on critical care to protect one from death is a false one. If preventive measures that would save more lives are forgone in order to provide critical care, one's life is *not* as well protected as it might have been. Thus, some of the possible benefits of using critical care as a symbol for life's pricelessness can be gained only by sacrificing the important values of honesty and openness in deciding how to allocate public resources. These choices cannot be faced openly without recognizing that lives will be lost whatever we do.

Claims for the special symbolic value of critical care are based on its well-known and sometimes dramatic ability to save individual lives. Loss of life due to failure to prevent hazards or diseases may not, however, always be invisible or symbolically impotent. The burned-out Pinto, for example, has in this country become a symbol for corporate and government decisions to forego safety precautions. An organization called Mothers Against Drunk Drivers (MADD) symbolically reinforces the life-saving goals and the urgency of its campaign against drunken driving by the fact that many of its members have had a child killed by a drunk driver. In at least some cases, then, loss of life due to lack of prevention may come to have symbolic representation and to be publicly recognized as the result of decisions not to protect or place a high value on life.

A *second* argument in support of critical care is based on the *physician's commitment* to individual patient welfare. This is the reason behind Del Guercio's plea that physicians not be forced to choose which patients are worthy of critical care, lest patients come to fear their physicians. Pellegrino has also pointed out that the patient expects his or her physician to act in the patient's own best interests, not those of distant patients, society or the common good ([34], p. 9). This expectation is a fundamental reason for granting special authority to the physician to act on behalf of his or her patients in a fiduciary relationship [31]. It is similarly held to be essential to the trust placed by the patient in his physician and thus to the therapeutic benefits which flow from an open and trusting physician-patient relationship ([18], p. 33). If, as Del Guercio worries, the physician's ability to care for critically ill patients is severely compromised by a lack of resources, he will no longer be able to represent the best interests of individual patients, but will be forced to become a kind of triage officer, weighing the needs of patients against one another and refusing to care for some.

This argument is based on the advantages for medical care of the physician's unwavering dedication to the welfare of each individual patient. Although its proponents do not present hard data to demonstrate these advantages, it seems obvious that a physician who is more committed to his patient's best interests will provide better care than one who is less committed. Moreover, the patient's belief in a physician's pledge to help him or her should encourage openness, reassurance and hope, all of which may have important therapeutic value.

It is not clear, however, that the physician's commitment to individual patient welfare is necessarily incompatible with limitations in the provision of critical care. As several authors [29, 45] have pointed out, if specific limitations on the provision of certain kinds of care are chosen and prescribed as a matter of public or institutional *policy,* then physicians will not have to make these decisions for individual patients. Physicians may, therefore maintain their dedication to the interests of individual patients within limitations which are not under their control. Although this approach to limiting availability of critical care may protect their commitment to individual patient welfare, physicians will likely object that the generality and inflexibility of administrative restrictions will undermine their ability to tailor therapy to individual needs, and thus cause needless harm to patients. Others, however, will respond that although formal restrictions may occasionally interfere with an individual patient's best interests, such restrictions are needed to correct practices which are widespread but ineffective.

This argument for critical care seeks to protect the patient's trust in the physician's commitment to his or her welfare. For some, however, this trust has already been destroyed by other factors; such persons question whether physicians have upheld an allegiance to the good of each of their patients, or at least convinced their patients of their benevolent intentions. Reports and rumors of great financial gain, exploitation, and insensitivity have inspired in some persons an attitude not of trust, but of suspicion regarding the physician and orthodox medical care. These individuals, then, might welcome limitations in the scope of the physician's power and a greater emphasis on self-care and preventive measures not provided by physicians.

A *third* argument for giving priority to critical care appeals to the *urgency* of critical care needs. Critical care is, I have said, the care of patients with severe, life-threatening illness or injury. Although what constitutes a need for medical care is, in most cases, not clearly defined and subject to wide variation in the judgments of individual physicians ([10], p. 20), life-threatening conditions surely comprise the most unambiguous, most pressing needs for medical care. The immediate, unquestioned response of most people, and perhaps especially health care professionals, is to give priority to those whose needs are most urgent. Can this impulse to help those who confront us with serious medical problems justify giving priority to critical care, or is it merely an emotional bias which we as a society must try to overcome in order to promote the health of all citizens most efficiently? Let us consider three arguments offered by Freedman in defense of what he calls "the imperatives of the present," the intuitive preference for those in immediate need of care ([18], pp. 36-37). First, Freedman claims that the benefits of preventive medicine are difficult to calculate, uncertain, and perhaps exaggerated. He makes reference here to Dubos' claim that new diseases would move in to fill the gap left by prevention of current problems. Freedman admits, however, that this argument cannot bear much weight, since we have nothing better to guide us in health planning than careful evaluation of past experience. Moreover, the value of critical care is also often uncertain; indeed, some would claim that physicians tend to exaggerate its potential benefits in recommending critical care for patients *in extremis*.

Freedman's second argument appeals to the role of the physician. Physicians, he argues, are directly responsible for their own actions, but not for the actions of their patients. Thus, the physician has a greater responsibility to treat a serious treatable condition than to get his or her patients to change their unhealthy behaviors. This is because the patients themselves ultimately choose to engage in risky or unhealthy actions. Since they cannot control patient behavior, physicians should not be held directlly responsible for it. Though this argument may explain why physicians should put greater emphasis on critical care than on prevention, it does not show that society should follow physicians in giving priority to critical care over other areas of health care.

Freedman's third argument turns on a distinction between saving lives and preserving health:

> ... there is a difference between saving the life of a person and preserving a person's health, even if that preservation will have the consequence, in some contemplated future time, of staving off death. If $10,000 can save one persons's life or preserve the health of two, which will have the consequence that they shall die later than they would have otherwise (at age seventy-five instead of fifty-five), the choice is not one of saving one person's life or saving the lives of two. (Two what? Fifty-five-year-old men? But the men are twenty-five now!) The translation does not go through. The choice remains what it was at the outset: that of saving one life or preserving the health of two men ([18], pp. 36-37).

This distinction between saving lives and preserving health is significant, Freedman argues, since saving a life is morally better than preserving health. He offers the following analogy:

> Which is worse: killing an individual today, or giving twenty people their first cigarettes today, when statistics indicate that three of them will die prematurely from the ill effects of smoking? ([18], p. 37).

Freedman asserts (and I am willing to grant) that in this example the former act, killing someone, would be worse than the latter, inviting or inducing disease. He claims that if killing is worse than inducing disease, then by analogy saving a life must be better than preserving health. Freedman appeals here to a mathematical model of "moral equivalences" (if $-x < -y$ then $x > y$); I am not sure that this "moral mathematics" is justified[1], but again, I am willing to grant the conclusion to see where the argument takes us.

Supposing then, that Freedman's argument thus far is successful, what would it show? The argument concludes that an *individual moral agent* in the appropriate circumstances ought to prefer saving a life to preserving the health of several. Does the argument also show that our *social policies* ought to reflect this moral difference by favoring critical care over preventive medicine? The answer to this question is much less clear, since the state may have different priorities than individual persons. Freedman recognizes this and completes the argument as follows:

> Even if the state must be as concerned for future good as for present good, the individual doctor is not in the same moral position; and for the state to force the doctor to act as

> though its concerns were his as well would be a contravention of the moral division of labor. It is wrong for outside agencies to render impossible the fulfillment by individuals of their individual moral tasks ([18], p. 38).

This final move, however, is unsound. From the fact that it is morally better for a person to perform act *a* than act *b* it *may* follow that the state should not interfere with the performance of *a*. Surely, however, it cannot follow that that person can justifiably demand that the state provide him with everything he requires in order to perform *a* or that he has a right to such support. It would be impossible for society to respect such a right, since the same claim could be made on behalf of all those who provide needed or useful services. Respecting all such rights, however, would overwhelm even the most affluent society's resources. Thus, society cannot be obligated to provide physicians with everything they need to save lives, even if saving lives is morally preferable to preventing disease.

I have thus far examined three kinds of arguments which purport to show that we ought to give priority to critical care even if its potential contribution to health is less than that of other measures. These arguments are based on the special symbolic significance of critical care, the value of the physician's single-minded commitment to individual patients, and the urgency of critical care needs. I have identified what I believe to be serious shortcomings in all of these arguments. I submit, therefore, that these arguments have not made their case; we must look elsewhere if we are to establish the conclusion that priority ought to be given to critical care, efficient or not. I turn now to a briefer look at some of the challenges to the current priority on critical care.

II. OBJECTIONS TO CRITICAL CARE

In recent years, a number of commentators on Western health care systems have challenged the established priority in the allocation of resources given to critical care for acute illness [1, 24, 26, 27, 40]. Though their criticisms and prosposals differ in many and important ways, these writers basically agree that a shift in emphasis from critical care to preventive medicine, self-care, and/or basic primary care will make a greater overall contribution to health at a much lower cost. The arguments of two well-known writers, Ivan Illich and Thomas McKeown, can serve as examples of this position. Illich is the more radical of the two, arguing in *Medical Nemesis* for a basic shift away from physician-provided and controlled care to a heavy reliance on individual self care. According to Illich:

> The social commitment to provide all citizens with almost unlimited outputs from the medical system threatens to destroy the environmental and cultural conditions needed by people to live a life of constant autonomous healing. This trend must be recognized and reversed ([24], p. 6).

McKeown's criticisms focus more specifically on acute medical care, as in the following passage from *The Role of Medicine:*

> The treatment of established disease, although important for patients, does not usually restore them to a life of normal duration and quality; and the modern improvement in health was due to the prevention of disease rather than to treatment after it occurred.
>
> The conclusion to be drawn is that the achievements of the acute hospital do not justify the relative neglect of the majority of hospital patients who are not admitted ([27], p. 195).

These and other writers object to current levels of support for critical care on several grounds. Critical care is said to be often *ineffective,* that is, unable to save lives in many instances, as indicated, for example, by high mortality rates reported in several recent studies from specialized adult critical care units [4, 11, 12, 13, 33, 41, 44].[2] Some critics point out that it can do positive *harm* by, for example, unduly prolonging death and isolating the dying patient [21]. Even where critical care is able to save lives, as, for example, in treatment for end-stage renal disease, or, perhaps, coronary artery bypass surgery, it is argued that the *costs* are too great to justify the benefits realized. Let us examine these objections.

If it can be *shown* that a treatment is ineffective, or even harmful, there is clearly little or nothing to recommend it. However great one's need for rescue or cure, the physician is not obligated to attempt a cure where no effective treatment for one's condition is available. Indeed, I would argue that the physician's obligation to prevent harm should include withholding clearly ineffective treatments which are painful or invasive, in addition to providing emotional support and palliative treatment for the hopelessly ill or dying patient. Surely, however, physicians recognize that providing ineffective or harmful treatments is not in the patient's best interest; the greater use of orders not to resuscitate for patients who cannot benefit from further aggressive treatment reflects this awareness [16, 38]. I also venture to say that most supporters of critical care, including physicians, would agree that there exist some very expensive treatments (such as germ-free environments for persons with combined immune deficiency disease), which, though they could be beneficial, should not be supported because their cost is prohibitive (see e.g., [42], p. 567).

Thus, the basic claim that clearly ineffective, harmful and/or prohibitively expensive treatments should not be supported is, I believe, uncontroversial. The controversy, of course, begins when one attempts to state which procedures fit these descriptions and under what circumstances. Physicians and social scientists who have examined these questions emphasize the relative paucity of the data and the need for much more research on the costs and benefits of critical care ([8]; [10], p. 109; [28], p. 269; [15], pp. 1236-37; [42], p. 567). What data does exist seems to present a mixed picture. On the one hand, several recent studies have found that a small number of high cost patients consume a large proportion of hospital resources - in one study 13% of patients with the highest charges consumed as many resources as the remaining 87% of patients in each of five general hospital populations ([46], p. 997). As noted previously, other studies of adult medical and surgical intensive care have suggested that a large amount of critical care is devoted to those apparently least able to benefit from it, namely, the terminally ill. If this data can be generalized, it would indicate that critical care can make only a limited impact on reducing mortality rates. It may, however, still be difficult to predict *beforehand* which individuals can benefit from critical care; for this to be possible, accurate prognostic criteria would be needed for determining which patients cannot be saved.

On the other hand, even strong critics of clinical medicine like McKeown and Carlson recognize that some critical care, such as treatment of acute emergencies, is both effective and efficient and hence should be supported ([28], p. 269; [6], p. 223). Bendixen argues that though intensive care costs vary tremendously, for some groups of patients, such as patients with respiratory arrest due to barbiturate overdose, intensive care has a very favorable cost-benefit ratio ([2], pp. 375-376). Others have made strong claims for the benefits of neonatal and pediatric intensive care [3, 22, 25]. Griffin and Thomasma assert that intensive care for infants and children is highly effective - more than nine out of ten children treated will completely recover, that is, will have no residual disability ([22], p. 158). In sum, then, given the paucity and inconclusiveness of the existing data, no *general* conclusion regarding the cost-effectiveness of critical care is clearly warranted.

III. IS PREVENTION THE ANSWER?

McKeown [25] and others (see [36], p. 25) have, I believe, clearly shown that basic preventive measures are primarily responsible for the vast improvements in health in the developed countries over the last three centuries. I do not wish to dispute their findings. Rather, I will consider questions about what *further* benefits preventive measures can hope to accomplish in our society in the future. McKeown argues that preventable diseases may be separated into two broad categories, those associated with poverty and those associated with affluence ([27], pp. 82-87). Diseases associated with poverty, particularly infectious diseases, were predominant in the past and still predominate in the developing countries. Many of these could be controlled through well-tried measures like provision of food, better hygiene, population control, and control of vectors ([27], pp. 168-169). In the developing countries, therefore, established preventive measures do promise a significant improvement in health. In the developed countries, however, where the most effective preventive measures have already been implemented, most of the predominant diseases are diseases of affluence, caused by overeating, smoking, drinking, physical inactivity, etc. Much of the further benefit of preventive medicine and self-care in developed countries, therefore, will depend on changing or eliminating these health-threatening behaviors. Many of these behaviors, however, such as smoking, drinking, and the use of firearms and motorcycles, are highly prized by those who engage in them, and thus, anything more than voluntary choices to give up these activities has been and will be strongly resisted in the name of individual freedom of action. Even many strong supporters of preventive measures admit that health education campaigns designed to promote voluntary behavior changes have not been particularly successful ([9]; [17], p. 280; [28], p. 265; [39]). Freudenberg, for example, claims that ". . . the record of health education in getting people to stop smoking, lose weight, drive carefully, avoid unwanted pregnancies or exercise more is extremely disappointing. If well designed, carefully executed studies have difficulty changing small groups of volunteers, the potential impact of such strategies on the general population is minimal" ([19], pp. 375).

The results of one very large and expensive recent study of preventive measures for heart disease (the MRFIT study [32]) raises the further question whether we have even correctly identified the most important risk factors for heart disease. Despite reductions in smoking, cholesterol levels and blood pressure in the study group targeted for special intervention, that group showed a statistically insignificant, only slightly lower 7 year mortality rate from heart disease, and a slightly higher overall mortality rate than a matched control group. Examination of subgroups led the MRFIT investigatiors to speculate that the inconclusive mortality results may be due to a harmful effect of antihypertensive drugs on some of the subjects ([32], p. 1475). Nevertheless, an editorial in *The Lancet* offered the following estimation of the study: "The results prove nothing, and we must turn elsewhere to answer the question, Does prevention work?" ([43], p. 803).

Even if we have correctly identified the most important health-promoting behaviors, however, we are faced with a basic conflict between effective but coercive preventive health measures and respect for individual freedom of action. Our legal system has generally upheld the freedom to engage in unhealthy or risky behaviors, as long as such behaviors do not adversely affect others [36]. If we as a nation continue to protect the liberty to engage in and to promote widespread but risky activities, and if voluntary efforts to change health behaviors remain relatively ineffective, the need for treatment for serious illness and injury will continue to be significant and preventive medicine will have only a limited success. Our policies and programs will not be consistent with our needs if we choose to deemphasize the provision of critical care despite other social decisions which seem to assure the continuation of a steady need for such care.

My conclusion, then, is that the arguments for granting a higher priority to preventive medicine are no more successful than the arguments for maintaining the current priority on critical care. Conclusive evidence for the overall inefficiency of critical care as well as for the effectiveness and feasibility of more widespread preventive efforts is, I think, still wanting. Given the present state of the controversy, I suggest that it would take a leap of faith to embrace one position over another. There may, of course, be other arguments or other data which can guide our allocation choices more convincingly. Pence's paper in this volume [35] clearly illustrates the threat to our health care system posed by the increasing costs of care. In view of that threat, we literally cannot afford to abandon our efforts to determine the best uses within health care of our limited resources.

East Carolina University School of Medicine
Greenville, North Carolina

NOTES

[1] For a critical analysis of Freedman's argument here, see [30], pp. 166-168.

[2] Consider, for example, Cullen's comments based on a study of 226 unstable patients requiring intensive physician and nursing care: "Intensive care medicine is extraordinarily expensive, yet results in only a small number of patients surviving to a useful and productive life. Far more likely is death or survival with poor functional recovery. High-quality intensive care resulted in a one-year survival rate of 27%. Only 12% of the entire study group were fully recovered and functioning as productively one year later as they had prior to their critical illnesses. This was accomplished at a total cost of $3,232,647, with $617,710 (21%) consumed by a limited resource - blood and blood products" ([11], p. 213).

REFERENCES

[1] Ardell, D. B.: 1978, 'Holistic Health Planning', in *The Holistic Health Handbook,* And/Or Press, Berkeley, California.

[2] Bendixen, H. H.: 1977, 'The Cost of Intensive Care', in J. P. Bunker *et al.* (eds.), *Costs, Risks and Benefits of Surgery,* Oxford University Press, New York, pp. 372-384.

[3] Budetti, P. P. and McManus, P.: 1982, 'Assessing the Effectiveness of Neonatal Intensive Care', *Medical Care* **20,** 1027-1039.

[4] Byrick, R. J. *et al.:* 1980, 'Cost-effectiveness of Intensive Care for Respiratory Failure Patients', *Critical Care Medicine* **8,** 332-337.

[5] Calabresi, G. and Bobbitt, P.: 1978, *Tragic Choices,* W. W. Norton and Co., New York.

[6] Carlson, R. J.: 1975, *The End of Medicine,* Wiley, New York.

[7] Childress, J.: 1979, 'Priorities in the Allocation of Health Care Resources', *Soundings* **62,** 256-274.

[8] Cochrane, A. L.: 1972, *Effectiveness and Efficiency,* Nuffield Provincial Hospitals Trust, Oxford.

[9] Cohen, C. I., and Cohen, E. J.: 1978, 'Health Education: Panacea, Pernicious, or Pointless?' *New England Journal of Medicine* **299,** 718-720.

[10] Cooper, M. H.: 1975, *Rationing Health Care,* Halsted Press, New York.

[11] Cullen, D. J.: 1977, 'Results and Costs of Intensive Care', *Anesthesiology* **47,** 203-216.

[12] Cullen, D. J. *et al.:* 1976, 'Survival, Hospitalization Charges and Follow-up Results in Critically Ill Patients', *New England Journal of Medicine* **294,** 982-987.

[13] Davis, H. *et al.*: 1980, 'Prolonged Mechanically Assisted Ventilation: An Analysis of Outcome and Charges', *Journal of the American Medical Association* **243,** 43-45.

[14] Del Guercio, L. R. M.: 1977, 'Triage in Cold Blood', *Critical Care Medicine* **5,** 167-169.

[15] Enthoven, A. C.: 1978, 'Shattuck Lecture - Cutting Cost Without Cutting the Quality of Care', *New England Journal of Medicine* **298,** 1229-1237.

[16] Epstein, F.: 1979, 'Responsibility of the Physician in the Preservation of Life', *Archives of Internal Medicine* **139,** 919-920.
[17] Fielding, J. E.: 1978, 'Successes of Prevention', *Milbank Memorial Fund Quarterly* **56,** 274-302.
[18] Freedman, B.: 1977, 'The Case for Medical Care, Inefficient or Not', *Hastings Center Report* **7:** 2 (April), 31-39.
[19] Freudenberg, N.: 1979, 'Shaping the Future of Health Education: From Behavior Change to Social Change', *Health Education Monographs* **6,** 372-377.
[20] Fried, C.: 1970, *An Anatomy of Values,* Harvard University Press, Cambridge, Massachusetts.
[21] Fuchs, V. R.: 1976, 'A More Effective, Efficient and Equitable System', *Western Journal of Medicine* **125,** 3-5.
[22] Griffin, A. and Thomasma, D.: 1983, 'Triage and Critical Care of Children', *Theoretical Medicine* **4,** 155-163.
[23] Hoff, C.: 1982, 'When Public Policy Replaces Private Ethics', *Hastings Center Report* **12:** 4 (August), 13-14.
[24] Illich, I.: 1976, *Medical Nemesis,* Random House, New York.
[25] Kaufman, S. L. and Shepard, D. S.: 1982, 'Cost of Neonatal Intensive Care by Day of Stay', *Inquiry* **19,** 167-178.
[26] Lalonde, M.: 1974, *A New Perspective on the Health of Canadians,* Government of Canada, Ottawa.
[27] McKeown, T.: 1979, *The Role Of Medicine: Dream, Mirage or Nemesis,* Princeton University Press, Princeton.
[28] McKeown, T.: 1981, 'Medical Technology and Health Care', in M. Staum and D. Larsen (eds.), *Doctors, Patients and Society,* Wilfred Laurier University Press, Waterloo, Ontario, pp. 259-272.
[29] Mechanic, D.: 1978, 'Rationing Medical Care', *The Center Magazine* **11** (Sept.-Oct.), 22-27.
[30] Menzel, P. T.: 1983, *Medical Costs, Moral Choices,* Yale University Press, New Haven.
[31] Moskop, J. C.: 1981, 'The Nature and Limits of the Physician's Authority', in M. S. Staum and D. C. Larsen (eds.), *Doctors, Patients and Society,* Wilfrid Laurier University Press, Waterloo, Ontario, pp. 29-43.
[32] Multiple Risk Factor Intervention Trial Research Group: 1982, 'Multiple Risk Factor Intervention Trial', *Journal of the American Medical Association* **248,** 1465-1477.
[33] Nunn, J. F. *et al.:* 1979, 'Survival of Patients Ventilated in an Intensive Therapy Unit', *British Medical Journal* **1,** 1525-1527.
[34] Pellegrino, E.: 1978, 'Medical Morality and Medical Economics', *Hastings Center Report* **8:** 4 (August), 8-11.
[35] Pence, G.: 1985, 'Report of the President's 2003 Commission on the Fall of Medicine', in this volume, pp. 163-170.
[36] Pollard, M. R. and Brennan, J. T. Jr.: 1978, 'Disease Prevention and Health Promotion Initiatives', *Health Education Monographs* **6,** 211-222.
[37] Powles, J.: 1973, 'On the Limitations of Modern Medicine', *Science, Medicine and Man* **1,** 1-30.
[38] Rabkin, M. T. *et al.*: 1976, 'Orders Not To Resuscitate', *New England Journal of Medicine* **295,** 364-366.

[39] Robertson, L. S. *et al.*: 1974, 'A Controlled Study of the Effect of Television Messages on Safety Belt Use', *American Journal of Public Health* **64,** 1071-1080.
[40] Sidel, V. and Sidel, R.: 1977, *A Healthy State,* Pantheon Books, New York.
[41] Thibault, G. E., *et al.:* 1980, 'Medical Intensive Care: Indications, Interventions, and Outcomes', *New England Journal of Medicine* **302,** 938-942.
[42] Thompson, L.: 1982, 'Critical Care Tomorrow: Economics and Challenges', *Critical Care Medicine* **10,** 561-568.
[43] 'Trials of Coronary Heart Disease Prevention', *Lancet,* October 9, 1982 803-804.
[44] Turnbull, A. D. *et al.:* 1979, 'The Inverse Relationship Between Cost and Survival in the Critically Ill Cancer Patient', *Critical Care Medicine* **7,** 20-23.
[45] Veatch, R.: 1981, 'Federal Regulation of Medicine and Biomedical Research: Power, Authority, and Legitimacy', in S. F. Spicker *et al.* (eds.). *The Law-Medicine Relation: A Philosophical Exploration,* D. Reidel Publ. Co., Dordrecht, Holland, pp.75-91.
[46] Zook, C. J. and Moore F. D.: 1980, 'High-Cost Users of Medical Care', *New England Journal of Medicine* **302,** 996-1002.

GREGORY E. PENCE

REPORT OF THE PRESIDENT'S 2003 COMMISSION ON THE FALL OF MEDICINE *

I come before you today on a special mission from the President's Commission on the Fall of Medicine in the year of your future 2003 - a year highlighted by the remarkable achievement of time travel and the dismal events which occurred in medicine. Time travel works much like transubstantiation whereby spiritual essences are exchanged in physical bodies retaining their external qualities, and hence, although the commentator on Professor Moskop's interesting, careful paper [4] appears to be Gregory Pence, this is not so. Instead, I, Stewart Lawrence Randolph III, have replaced the soul of Professor Pence for this presentation. Why the Commission sent me at this particular time and place is not for me to question or understand; I shall accept the opportunity graciously.

In fairness, I should tell you that determinists and fatalists on the Commission view this mission with great doubt, but as you might infer, the freewillers are in the majority. The former think my mission will have no effect whatsoever on the future of medicine in your time.

Finally, do not worry for the moment about Professor Pence, for he is now suspended in metaphysical limbo and his soul will only return to his body when the last word of this commentary is finished. Of course, he will remember absolutely nothing of it, nor should you expect him to bear any responsibility whatsoever for the following remarks.

I

To help you understand how the Great Crash of Medicine occurred in 2003, I shall first discuss certain empirical trends which were apparent even in your own time of 1983, and then discuss the philosophical dilemma which helped cause the Great Crash. In this way, it will be obvious how the general issue discussed by Professor Moskop - critical care medicine and distributive justice - became *the* issue of the latter part of the twentieth century.

I believe it was about this time when your Congress attempted to correct the long-term, financial problems of Old Age, Survivors, Disability,

J. C. Moskop and L. Kopelman (eds.), Ethics and Critical Care Medicine, 163-170.

and Hospital Insurances (OASDHI). Many of you and your Congressmen focused only on the generational conflicts involving higher taxes or lower benefits, believing the myth that such a combination could restore the integrity of the system. Yet the Chief Actuary of the Social Security Administration repeatedly stressed that three fourths of Social Security's problems were due to the uncontrollable escalation of *health care payments* [7, 8, 9]. Over the next two decades, every attempt to address the real problem was rebuffed by alliances formed by the unions, the elderly, the poor, and public employees – all under the leadership of the AMA. The real implications were obvious.

Another ominous trend apparent even in 1983 was the great growth in Veterans Administration (VA) payments for medical services. In the twelve years between 1968 and 1980, expenditures for such services jumped 410% from 1.6 billion to 6.6 billion dollars [10]. Despite this, your Presidents Carter and Reagan enormously increased military personnel and their benefits in subsequent years. In 1983, a 100% disabled veteran received $2,000 tax free per month, plus free VA room and board for life. By 1990, half of such disabilities were psychological and cost $5,000 per month to maintain or $60,000 per year. Thus VA medical payments jumped 1000% over the next 12 years between 1983 and 1995.

The philosophers on the Commission maintain that your real difficulty in cutting costs stemmed from opposing, contradictory moral views which could not be reconciled in medical areas. As we shall see, they may be correct, but I feel compelled to emphasize that our country has *always* had such philosophical tensions. The philosophers, however, maintain that the crucial difference in your time was the unstoppable growth of government and its entitlement programs.

Wherever the truth lies, it is clear that philosophical differences fueled the debate in the latter years of the twentieth century over the burgeoning numbers of recipients of Aid to Families with Dependent Children (AFDC). In your own time, the 1980 census revealed an enormous jump over the past two decades from 3 million to 11 million AFDC recipients [10]. Certain politicians decried the continuing rise of such recipients in the 1980's while continuing to condemn abortion, contraception, and sex education. It was very common in these years to hear young workers and the elderly condemn the swelling public assistance rolls while simultaneously condemning abortion. The same people who screamed about skyrocketing medical premiums forced legislatures to pass bills mandating maximum medical care for all defective babies - no matter how deformed and no matter what the parents decided. Aided by this sanctity-of-life rhetoric which ignored finances, AFDC recipients jumped 410% over the next twenty years to 40 million in 2000. The number of "infants" in indefinite intensive care soared 5,000% between 1983 and 1990 alone.

Now let me say something about your previous successes in preventive medicine, It is amazing to some members of the Commission that it was thought by some in your time that preventive medicine would save money. In truth, it did save *you* money, but it also merely allowed those who formerly died quickly to live an average of twenty years longer while dying of expensive, chronic, lingering diseases like emphysema and cancer. Moreover, as the average life approached 80 by the year 2000, your so-called "baby boom" generation reached their mid-fifties and began massively to drain medical insurance funds. In essence, the monies saved by prevention in the 1970's were monies borrowed against the future - and the debts came due at a very unfortunate time.

Some members of the Commission, the economists and social scientists, blame the radical changes which occurred in medicine in the 1980's. By 1985, the number of physicians had swelled enormously from 20 years before, and the proportional, usual, huge increase in medical services performed followed with almost scientific precision. Supply and demand have always not only failed to operate in medicine, but operated inversely: more surgeons result in more surgery per capita being performed at a higher cost per patient. Moreover, the way you *financed* medical growth directly contributed - in the eyes of medical economists - to the ultimate financial collapse of medicine. Hospitals mushroomed up everywhere in the 70's and 80's financed by long-term bonds bearing interest rates of 8 to 12%. The average medical graduate of 1983 was $21,000 in debt; by 1990, this figure rose to $40,000 at 12% interest [1]. *Interest payments and debt retirement alone* for medical financing raised medical costs 5% each year, every year, during the last decades of your century. As with social security and military benefits, you had created benefits for your contemporaries to be paid for in large part by those who came after and who could not protest your decisions. Such was the burden you left to your children and grandchildren.

In addition to these changes in medicine, the rise of for-profit hospitals and nursing home chains skimmed the cream of private insurance funds, leaving public hospitals bankrupt and more and more often forced to beg from legislatures at a time when every lobbyist was also on his knees. Also, the great increase in allied health personnel continued as you hired and graduated more and more nurses, physical therapists, radiation technicians, psychologists, and lab technicians. By the year 2000, one in five people worked in some way in the health care sector. I even understand you had *philosophers* on payrolls in some of your medical schools, as well as ministers in your hospitals.

Even in your year of 1983, some of you must have begun to see the consequences of these trends for the average worker. By the year 2000, the tax burden on the average worker was overwhelming as more and more was taken from him and her to pay for old-age pensions, personal hospital care, veterans, AFDC recipients, and physicians' fees. Of course there was no doubt that medical insurance premiums were forms of taxes, especially since many workers had no choice about whether to purchase such insurance. By 2000, the average worker had 70% of his pay withheld for taxes and medical insurance premiums - withholdings which drove many workers to the more radical political movements which grew during the early years of the twenty-first century.

The problem for the worker was even more obvious for that class of *un*employed, *under*employed, and *marginally* employed workers who became a permanent fixture of American life over the next decades. I even remember a case years ago in my home state of Alabama – about the year of 1983, I believe – which showed the pattern of future events in medicine. The city of Mobile during this year suffered a 16% unemployment rate and 50% of patients presenting at Mobile Hospital's emergency room had no money and didn't qualify for any insurance [2]. Subsequently, this hospital – the only major-trauma facility for the surrounding 13-country area – closed its emergency room doors from 5 PM Friday to 8 AM Monday morning. Where victims of Saturday night gunshot wounds went, we do no know. The analogous, public, charity hospital in Birmingham at the same time was in serious financial trouble. It is ironic that in Birmingham at the same time, not one, but two, heart transplant programs were beginning at costs of over $100,000 per patient.

For these reasons, the question of preventive care in medicine versus critical care was becoming *moot,* even in your own time. The real question was *not* whether critical care might be wasted in spending 80% of certain monies on the last year of life of 13% of the population, nor was the real question whether preventive care might save money. Instead, the real question was rapidly becoming: *critical care for whom?* Would everyone get minimal critical care for acute, life-threatening emergencies, would programs be considered, or would only those get care who were covered by some insurance?

As these pressures mounted, society looked to individual physicians to contain costs. This proved impossible. Once a physician had accepted a certain patient and his family, the physician usually felt uncomfortable denying care when he knew of other ways to obtain that care. Even if he received no payment, physicians often got their patients care by bending certain rules, passing on costs to the nameless, faceless "society at large."

A more profound problem for physicians in denying care was seen in the quotation mentioned by Professor Moskop from Dr. Del Guercio at the opening of his paper. Unable to make such decisions itself, society called upon physicians to become *moral judges* of who did and who did not deserve medical treatment. This was tried briefly, but it proved fruitless. In essence, society had demanded contradictory sets of virtues (excellences) from physicians. On one hand, society expected physicians to have the virtues of compassion and fidelity to the individual patient. On the other hand, society expected physicians to be good judges, where good judges require the virtues of strict impartiality, allegiance to the public good, and a brutal honesty about compassion which often conflicts with true compassion. Individual physicians in the late 1980's and thereafter found themselves in an impossible dilemma of increasing intensity.

I must report also that some cynics on the Commission - with whom, of course, I disagree - attributed the rise in costs to conflicts of interest among individual physicians, for in limiting medical services drastically, a physician also limited his income. Certainly the American Medical Association worsened matters with its vigorous opposition to any real controls over medical decisions or to any cuts in government medical payments.

The national media also did not help much. While constantly mentioning the rise in medical costs, the media continued to emphasize dramatic cases where children would die for lack of a certain treatment. In each case, some hospital or government agency finally gave in, thus encouraging more families to seek media attention.

At this point, I must report the views of the philosophers on the Commission mentioned earlier who see the roots of the Great Crash not in financial matters, but in a *moral impasse* you had reached in 1983. The two symbols of this conflict – a conflict essentially about distributive justice in medicine – of course were your two philosophers, John Rawls and Robert Nozick. These philosophers and their great books articulated in a more precise way the rough divisions and polarization of your society in 1983 [5, 6]. For Rawls, just allocation in medicine minimized the natural inequalities of health, wealth, and talent which Fate had decreed. Instead of "trickle-downs," Rawls favored *flooding-up* from the bottom. For Nozick, just allocations in medicine were just the opposite, whereby the rich, the intelligent, the healthy, and the motivated were free to lead their own lives unfettered by interference from the government. This seemed a reaction to a kind of taxation-slavery which was beginning to occur in 1983 in an attempt to foster Rawlsian ideals of justice for medical unfortunates. Nozick championed rights to property and the individual liberties of both physicians and the healthy to be left alone; Rawls in essence institutionalized *agape* in medical institutions. When these two philosophic conceptions of justice entered the world of medical finance, we saw a straight contradiction: a raw, 19th century, classical liberalism running up squarely in the face of the Welfare State – individual liberties facing off against institutionalized charity.

In his 1981 book *After Virtue,* Alasdair MacIntyre had argued that these two conceptions of justice were not just incompatible, but also *incommensurable [63]. They were philosophical starting-points* at a crossroads leading in opposite directions, such that travelers on one road could not recognize anything in common with those on the other road. As such, travelers on the two roads had opposing definitions not only of just allocations in medicine, but also of charity and greed. Most profoundly of all, travelers lacked agreement not only about those shared moral rules which turn the jungle into civilization, but also about answers to the ultimate question: What is society *for?* What is the purpose of medicine?

These incompatible conceptions of society and justice were not merely found in the fanatical champions of each side; they were instead found in every citizen, every physician. As we see it from the hindsight of the future, your society had inherited incompatible moral traditions and their incompatibility was seen every day in medicine. Using Rawlsian-like views, many thought it was *unfair* that a 12-year child could not have a heart transplant by taxing healthy workers a few pennies more each week; in contrast Nozickians argued it was not *unfair* that the poor 12-year old could not otherwise have the transplant. Indeed, for Nozickians, the idea behind such taxation, generalized thousands of times to cover millions of people, amounted to immoral taxation-slavery of healthy workers. For Nozickians, the only unfairness would have been if parents and a surgeon had entered a contract for the child's surgery and government prevented its execution. Otherwise, for Nozickians (and a medical philosopher named Engelhardt) it was not *per se unfair* that some were born wealthy, others poor – some healthy, others sick. For Nozickians, that was just life; for Rawls and his followers that was what was *wrong* with life, something to be addressed by a just society.

By the mid-1990's, questions of critical care and its allocation had become revolutionary. The year 1993 saw the beginning of the despicable practice of kidnapping children of physicians by poor parents who were desperate for operations for their children. Such terrorists argued that if their own children couldn't live, neither could those of surgeons. The same year witnessed the terrorist bombings of the for-profit hospitals which continued for the next decade.

Somehow, society held together and hobbled along to the year 2003. Although citizens labored under taxation which would have been intolerable in 1983, somehow people adjusted. Many prayed for a miraculous solution.

A miraculous solution did not occur, but a tragic one did. In the year 2003, government and private medical insurance funds declared themselves broke. By then, the two kinds of insurance funds were inextricably linked together. Strong governmental controls were immediately imposed. Physicians attempting to leave the country found their bank accounts frozen. Several were stopped at American borders with large quantities of gold.

We, the citizens of 2003, have attempted to revive private and public medical insurance, but there is as much resistance to that as to the new paper money. Many physicians now are bitter about their new-found poverty. Medical services are rendered only after prior payment, usually in precious metals. Few people can pay much.

There are two bright spots in our current picture with which I would like to leave you. First, more than enough students still opt for medicine, although we have drastically reduced the number of schools and places for medical students. These students tend to be idealistic intellectuals who are committed to a new vision of scientific medicine. They seem to be the wave of the future. Second, despite the doubts of the sceptics, we have seen the revival among middle-aged physicians of the ancient virtue of charity. Nor is this the old, pseudo-virtue of charity where a physician treated a poor patient but expected society to pay him. No, in the future we have seen that charity which is a free, spontaneous giving of the healer from his heart to those in need. Perhaps, with these new students and this noble virtue among our older physicians, we are not so bad off after all.

University of Alabama in Birmingham
Brimingham, Alabama

NOTE

* For obvious reasons, Professor Pence's commentary was not altered from the lecture format. Eds.

BIBLIOGRAPHY

[1] Boerner, R.: 1983, Editorial, *Journal of Medical Education* **58,** 669-670.
[2] Loeb, M.: 1983, 'Rising Joblessness Forces Hospital in Alabama to Trim Emergency Care', *Wall Street Journal,* Feb. 10, 29, 38.
[3] MacIntyre, A.: 1981, *After Virtue,* University of Notre Dame Press, Notre Dame, Indiana, pp. 227-238.
[4] Moskop, J. C.: 1985, 'Allocating Resources within Health Care: Critical Care vs. Prevention', in this volume, pp. 147-161.
[5] Nozick, R.: 1975, *Anarchy, State, Utopia,* Basic Books, New York.
[6] Rawls, J.: 1971, *A Theory of Justice,* Harvard University Press, Cambridge, Massachusetts.
[7] Loeb, M.: 1983, 'Rising Joblessness Forces Hospital in Alabama to Trim Emergency Care', *Wall Street Journal*, Feb. 10, 29,38.
[8] Peterson, P.: 1983, 'A Reply to Critics,' *New York Review of Books* **30:** 4, 48-56.
[9] Robertson, A. H.: 1982, *The Coming Revolution in Social Security,* Reston Publ. Co., Reston, Virginia.
[10] U.S., Department of Commerce, Bureau of Census: 1982, *Statistical Abstract of the United States,* U.S. Government Printing Office, Washington D.C., pp 320-374.

JOSEPH MARGOLIS

TRIAGE AND CRITICAL CARE

Medical triage is a much muddled matter. One surprisingly important reason that it is, has to do with the sorting of wool and coffee. The use of the term *triage* is noted, in the *OED*, for the sorting of wool fleece as early as 1727 and for coffee in 1825. The specimen entry for coffee indicates that beans were standardly graded as "best quality" or "middling," and the third sort, the so-called "bad broken berries," were called "triage coffee." The *OED* also notes that the verb *try*, which comes from the same root, *trier*, to cull or sort, had already acquired its legal use in Anglo-French practice *circa* 1280. But it seems to have been used then, as in a way it still is, to signify distinguishing wrong from right – which of course by a not unreasonable extension could be thought to bear on triage of the market sort, in the sense that one might falsely represent one grade of wool or coffee as another; but this would still be to mix two distinct ideas. The important thing is that the earlier wool sorting does not seem to have featured any tripartite scheme and is not associated with any market emergency; and the triple division of coffee beans, perhaps etymologically innocent, has very noticeably yielded to an almost irresistible three-fold classification in modern uses of the notion of triage – as in emergency military medicine. Military medicine is not invariable in this; but certainly one of the most widely cited schemes for either transporting the wounded for medical treatment or actually treating them divides the pertinent population into those who are too badly hurt to benefit at all, those who stand a reasonable chance of recovering without any intervention, and those whose prognosis is clearly maximally or decisively affected by the kind of intervention being considered.[1]

There are enormous differences between the practice of coffee and military triage. But strange as it may seem these differences have never been completely sorted in a careful way. The result is that certain pressing questions regarding newer forms of triage – notably with respect to the technological advances of modern medicine – are very nearly reduced to incoherence. There is, first of all, no attention to scarcity or shortage or crisis or need of any sort in speaking of coffee triage; practice, there, has to do, it seems, exclusively with the grading of commodities in

J. C. Moskop and L. Kopelman (eds.), Ethics and Critical Care Medicine, 171-189.

accord with market norms – that are themselves of course linked to prevailing tastes and demands.[2] If scarcity has any relevance at all in the market practice, it is as a *result* of triage, not as a moral or legal or prudential problem of some sort to which triage is applied as an appropriate and generally *adequate method of resolution*; or, further afield, scarcity is a mere contingency on which the practice has an entirely tangential bearing. The distinction between features of the case to be decided and features of the decision itself will prove essential (as we shall see).

Secondly, it would be excessive to suggest that the specific satisfactions of graded wools and coffees belong to any list of human needs that could be seriously said to be essential or to take precedence over such values as actual survival, avoidance of disability or profound pain, or the like (although in the case of morning coffee perhaps we might relent). But it is clear that one might well speak of a fair practice with regard to sorting coffees and, consequently, with regard to pricing and related distributional considerations concerned with justice in the market. It would then also not be unreasonable to raise higher-order questions about whether the presumed justice of practices linked with coffee triage – for example, pricing and availability within a given market or with regard to the very scope of the market to be served – accord or fail to accord with utilitarian or contractarian or egalitarian principles or the like. But although it would be possible to institute a top-down practice of justice in which triage itself could be said to be *entailed* by some would-be utilitarian criteria by which the market served the greatest good of the greatest number, it seems much more persuasive to suppose that coffee triage was a natural, somewhat local practice with only the thinnest connection with such universal principles and that the attempt to apply such principles in the actual working practice might well generate alien and utterly unmanageable problems. The trouble is that it is just this conception that is more often than not invoked in tendering paradigms of medical triage. Thus, for example, Stuart Hinds emphatically reports the decision, in 1952, of the British Ministry of Health – faced with a short supply of Salk polio vaccine and the imminent danger of a high death rate among unvaccinated children – that to make the supply "go round most equitably, all eligible children would have their names listed for a lottery-type selection... . This is 'triage' [he adds]" ([16] p. 39). But either this is *not* triage or not the only or most plausible interpretation of triage; or the point of the lottery, open to a crucial equivocation, is never explained in terms of triage itself.

Now, medical triage is *not* market triage applied to medicine – or, perhaps better, if it is only so construed, then it cannot capture the kinds of conceptual difficulty that recommend its close analysis in moral terms. For surely medical triage, whether during battle or in the not so metaphoric wars against human misery, is essentially concerned with: (i) scarcity or shortage of goods and services of a medically considerable sort; (ii) goods and services publicly perceived to be available in some measure and expected, or open to request, in accord with well-defined practices or their extension (because, say, of technological innovation); (iii) resources in demand here and now or in some reasonably anticipated immediate short run, and outstripping present supplies; (iv) resources designed to provide substantial relief with regard to relatively fundamental needs and desires (as by saving life, preserving health, reducing suffering, and the like). So seen, it is easy to imagine that the notion of medical triage could foster parallel applications in disasters and crises of other sorts – for instance, in managing famine relief, earthquake victims, the poverty-stricken, the jobless, and those lacking educational facilities at least. In this sense, triage cannot fail to bear on all the misfortunes of the human race.[3] We must, however, guard against too easy an enlargement of the range of triage. The core considerations are the ones mentioned; and the theoretical importance of the paradigms of triage is directly related to their remaining reasonably well-demarcated. In fact, the entire point of resisting generalizations that smooth out the differences between clusters of distinctive cases is just to draw attention to the dubious classificatory and theorizing tendencies that now dominate moral philosophy.

Nevertheless, to sort victims or patients or clients in medical contexts constrained by conditions (i)-(iv) – *on any comprehensive principle of ordering whatsoever* – is not, or at least not yet, to practice medical triage. Or, if it is, then medical triage, however humanely construed, is hardly yet more than a sort of moral coffee triage. In point of fact, it is often so construed. The *OED*, for instance, offers in its 1971 Supplement a specimen sentence from 1930 that introduces the use of the term *triage* to signify a "sorting station," at which someone (a "triage officer," for instance) examines soldiers, as under battle conditions, "to determine the urgency of their injuries." Nothing more and nothing less.

Blakiston's Gould Medical Dictionary goes further but is still inadequate [2]. The entry reads:

> *In military medicine,* the process of sorting sick and wounded on the basis of urgency and type of condition presented, so that they can be properly routed to medical installations appropriately situated and equipped.

There is no mention of scarcity at all; in fact, the entry gives almost the opposite impression. *Stedman's Medical Dictionary* gives a fuller entry, extending triage to "military or civilian disaster medical care" and specifically indicates division "into three groups: those who cannot be expected to survive even with treatment, those who will recover without treatment, and the priority group of those who need treatment in order to survive" [26]. But the rationale is entirely omitted, which makes the indicated sorting either pointless or unclear.

There is no need to quarrel about definitions. What is absolutely crucial for medical triage (on the interpretation, frankly, here favored) is this condition: there exists a disproportionate need and demand for immediate or imminent care or treatment in the face of scarce supplies and services *and,* because of that, the ordering of those to be served can be managed only with the understanding *that not all those eligible can be served and that that remains an evil that cannot be adequately justified by whatever covering policy justifies the treatment or care of those thus treated and cared for while omitting the others.* This is what (understandably) is missing from the *OED*, from Gould's and Stedman's dictionaries – and (less defensibly) from a frighteningly large number of discussions of the triage problem.

The triage problem *can* be completely eliminated or brought into line very nicely by construing it as requiring no more than a form of (moral) coffee triage. There will be those of course so impressed with the tenacity of ecological evil that they will not be persuaded by either of these solutions (elimination or reduction). Furthermore, *if* we resist the alternatives mentioned, it will turn out that some rather remarkably powerful consequences follow at once from the quite elementary distinction just introduced (treating some and omitting others). For example, it would suddenly become utterly impossible to resolve medical triage problems *as such* by an appeal to *any* egalitarian principle. The reason is clear: such a principle would require that *all* who are eligible are equally eligible (or, more strongly, that all [in the relevant context, that context not prejudicially construed] *are* equally eligible); hence, the egalitarian principle applied to all, so that only some are served, *cannot* (in the triage context) *justify* denying treatment or care to the others. *No* lottery, no random selection, *can* meet the requirements of the triage problem if the condition stated above is acknowledged; for, morally construed, a lottery principle implies that, in the circumstances *given*, the lottery will yield the best or fairest or (at least) a reasonably good and fair resolution *of the entire ordering problem*. But in that case, the resolution will have had to construe the triage problem as a problem of an altogether different sort or as nothing more than a moral form of the coffee triage problem (where, that is, all are equally entitled to be recognized for the grade of coffee bean they are). Egalitarianism or an egalitarian lottery can be read either way – which is instructive – but it cannot be read in terms that yield a solution to the triage problem construed in the *sui generis* manner we have supplied. Triage disallows what egalitarian lotteries justify – and does so trivially.

Put another way, it *is* possible to formulate an egalitarian principle of justice that accommodates scarcity and yet does *not* distribute scarce goods and services equally to all in need or to all entitled. We could for instance distribute scarce artificial hearts to those in need, by lottery, in such a way that not all would be accommodated and yet all would have been equally considered. Apart from whether that would be a defensible form of egalitarianism,[4] such a policy would presuppose that the apparent evil of the triage dilemma was only an *initial condition* set for a good solution all in all, not an ineliminable property of the *solution itself*. Perhaps we can read the problem either way, but we cannot suppose that the two readings are equivalent.

Medical triage, therefore, poses the uncomfortable problem of how to make a rational (and moral) choice in circumstances in which evil is (here and now) ineliminable *and* in which provisionally eliminable evil remains as a *result* of the actual choice one makes. It is not part of the triage problem that the initial evil be the work (however indirect) of human agents; although, as in battle, it is often bound to be. Triage does not arise in circumstances of utter incapacity or where men are driven to the most desperate expedients for survival (Hume's problem about the limited pertinence of justice [17]). This is why it is a mistake to construe Garrett Hardin's so-called "lifeboat ethics" (linked, for instance, to what Hardin calls "the tragedy of the commons") as a form, or the paramount form, of triage ethics. Or for that matter, the resolution of Malthus's prophecy. Undoubtedly, there *are* triage problems that arise from *here to* the threatened extinction of a given society; but decisions about the bare survival of a society *collectively* construed are distinct from and, presumably, take precedence over triage – which is itself premissed on the continuing survival of that society and (within that condition) on relatively moderate (though immediately disproportionate) scarcity [12, 13, 14, 15, 8]. Triage arises where less than adequate means are available for the survival or well-being of all or of all pertinently eligible; *where the distribution of relevant resources lies in the hands of responsible persons;* where those providing services are *not* among those who are to benefit; and where *any* of the policies that might be favored will knowingly fail to address the plight of some who are eligible (variably, for different policies).

In this sense, triage problems are either special forms of the famous moral dilemma cases or analogues of them. The difference perhaps is that moral dilemmas (like that of being able to save the life of only one of two strangers) are often thought to be fairly solved *by* lottery; whereas, in triage cases, this is either rejected or taken to be strongly contestable – or else open to an interpretation quite different from that of (what we may call) a *principled* lottery. The reasons are instructive. Triage presupposes technically or professionally developed provisions and services prepared in anticipation of needs and demands of a relatively fundamental kind, authorized somehow in the spirit of a society's concern, so considered by potential recipients, intended in due time and within a given society's capacity to be as adequate for impartial use as possible, normally not distributed so as to benefit the donors or those who decide the ordering arrangements, but confronted here and now by a crisis or legitimate demand disproportionate to its capacity to respond. By contrast, in many moral dilemma cases – particularly in so-called lifeboat cases – those who take charge of distributing resources or assigning obligations under conditions of dire scarcity or emergency often stand to benefit considerably by their own decisions, which cannot fail to affect fairness and impartiality.[5] Triage does not arise here, though the ordering of priorities relative to extreme scarcity and extreme need may well mislead us. It is also *not* a triage solution to distribute smaller but probably adequate doses of a life-saving drug in short supply to five patients instead of administering a massive but absolutely secure dose to one; it is *not* a triage problem that, having exhausted the available drug for a first, lone patient, no more remains for the subsequent needs of five new patients. But it *is* a decision presupposed by the triage procedure that a life-saving drug in short supply should be administered to patients in need, in spite of the fact that it cannot be supplied to all in need or to all those who are legitimately desirous;[6] and it *is* a triage decision to withhold scarce drugs from those now in need, anticipating the imminent need of others. The matter is obviously a vexed one.

If our account holds, then it will hold as well against all forms of utilitarianism. For, whatever else may be said of them, utilitarian criteria make sense only if they are applied in a principled and justified way to the *total* population affected. But the point of the triage problem is, precisely, that, on *any* such principle, the results will insure a measure of remaining evil that could and (arguably) should have been reduced or suitably altered or eliminated on *some* policy (if any policy were both effective for some part of the population in question and normatively involved). Here, the entitlement to care of those who could have been assisted *but were not* cannot be overridden, *on our interpretation of triage,* by any policy that claims to benefit, maximize, or optimize *the needs and legitimate demands of a given population as a whole* – in either the aggregative or the collective sense (should anyone construe utilitarian concerns as equivocal in this regard).[7] In short, troublesome as it may be, the triage orientation is opposed to construing the pertinent needs and desires of *individual persons* as invariably subject to overriding aggregative or collective moral judgments – even though it itself employs general ordering criteria in the attempt to rationalize the distribution of relevant goods and services under the condition of disproportionate emergency need. In this sense, triage cannot, as such, be subsumed under the principles of egalitarian or utilitarian justice; although whatever justice it may presume to dispense (if indeed it can claim to dispense any justice at all) can in principle be easily displaced – for instance, simply by asserting that its policies can be overridden by egalitarian or utilitarian policies *or* by construing the latter as themselves constituting effective triage policies.

These charges deserve to be made clearer. Let us speak of a *principled* ethic as one that judges particular cases under principles – not merely in the formal sense of consistency ("similar cases must be similarly judged in similar circumstances") but substantively (so that particular cases are to be judged *only and always* as instances of some general sort or sorts taken, in some prior sense, to be morally or legally or similarly significant). And let us speak of an ethical judgment as a *verdict* if it yields an overriding assessment of goodness, rightness, obligation or the like that rests on what are taken to be *all* the pertinent kinds of principled considerations (without knowingly excluding any such considerations and without necessarily, of course, being able to engage all pertinent considerations). We need not insist, then, that all principled ethical judgments are verdicts; although, on the thesis, we must hold that, if valid, such judgments contribute essentially (possibly in a variety of ways) to the formulation of bona fide verdicts. Without subscribing to their moral intuitionism, then, the distinction between principled ethical judgment and verdict corresponds more or less to what C. D. Broad intends between "component fittingnesses" and "resultant fittingness" and W. D. Ross, between "prima facie" and "overriding" claims [3, 25, 18]. In a principled ethic, normative predicates are applied to human beings, their behavior, character, dispositions, traits and the like only in virtue of the way these instantiate general categories of ethically sensitive concern over which the principles range. The ethics of triage, however, is indifferent to this distinction, since (on the account here provided) it opposes all forms of principled ethics – that is, opposes their adequacy for relevant cases, not the admissibility or pertinence of principled judgments otherwise. The thesis, then, is that: (1) not all would-be verdicts can be verdicts; (2) principled judgments that are not verdicts cannot always form part of a principled ethic; and (3) in some circumstances, it would be inappropriate to construe the relevant judgments as principled even if they were not verdicts. *Triage judgments are specimen judgments that meet these three constraints.*

This is simply the formal meaning of acknowledging that, in triage, the *solutions* tendered actually yield a perceptible evil (eliminable on some alternative possible distribution) that cannot be overridden by putatively relevant verdicts *or* by non-verdict-like principled judgments. This is precisely what triage problems challenge regarding the competence of egalitarian and utilitarian principles. One might add, it also features what is so worrisome about the verdictive optimism of so-called impartial lotteries and utilitarian sacrifices. In a word, triage tends to support very strongly a theory that holds that the evil of pertinent human crises cannot be rationally discharged by principled ethical judgments. It is an interesting fact, in this regard, that, just in addressing the problem of "ELT" (of allocating "exotic medical life saving therapy"), Nicholas Rescher, for one, straightforwardly affirms that "Moral philosophers of the present day are pretty well in consensus that the justification of human actions is to be sought largely and primarily – if not exclusively – in the principles of utility and of justice" ([24], p. 175). He sees the allocation problem as a triage problem and he resolves it by a lexically ordered pair of criteria ("inclusion" and "comparison") yielding verdicts

–by applying the first of which, as he says, we can justify "eliminating *en bloc* whole categories of potential candidates" ([24], p. 175). But this, arguably, is precisely what we cannot do.

Triage *does* invoke orderly distributional procedures matching scarce resources and present or imminent need – as emergency military medicine and the allocation of scarce drugs and transplants and the like make clear. Is there a contradiction here? The fact is that there is *no* distributional practice that could be validated by egalitarian or utilitarian principles (or by any other principle of a similar sort) that could not be consistently adopted in accord with the triage orientation. One could employ a lottery for instance in distributing scarce polio vaccine among children. The supporting reason would be, perhaps, that there simply is no way to sort a relatively healthy population of children in order to justify something like the classical tripartite ranking; and of course even that ranking could be interpreted in a principled way. Nevertheless, it would still be true that the triage justification was not a principled justification – in the sense favored by egalitarianism and utilitarianism – even though it could be said to accord with the "principle" of triage. This is bound to be puzzling.

The explanation is really quite straightforward. On the account sketched, a principled justification is one that maintains: (a) that the pertinent normative appraisal of a particular instance is arrived at by and only by construing that instance as a member of a certain ethically sensitive class of phenomena; and (b) that the putative evil of any state of affairs open to ethical resolution is itself the result *only* of failing to effect a principled solution or of failing to do so correctly. Now, triage is not incompatible with any rational rule for grading or ranking cases: it is only incompatible with construing doing so *as a* principled practice; and that, one may fairly say, is, for its defenders, "a matter of principle." The difference is an intensional one, since it is very likely that the distribution of polio vaccine, say, would be extensionally indistinguishable if practiced on the basis of an egalitarian lottery or practiced on the basis of moral triage that employed a form of fair lottery. This intensional difference may well make for motivational differences regarding future preparations of an ethically sensitive sort; but it also signifies, here and now, a radically different assessment of what one is accomplishing *in* distributing scarce polio vaccine to children in need, *without being able to serve them all.*

There is a deeper implication. A partisan of a principled morality might claim that that is all very well and good and very easy to say, but there simply is and can be no alternative basis for a *rational* ethic than a principled one. How is one to answer? Well, the proper response is to supply an alternative. There is only one – and it is more than an alternative. It is this: paradigm cases may be acknowledged in some consensual way, say, by inference and query from actual practice, and cases in question may be decided by comparison with paradigms (which may well change diachronically) within the ongoing practice of such decision. Morality – as well as much else regarding human concerns – may then be managed *case-by-case* rather than in a *principled* way (or better, principled judgments may themselves function as provisional and stable conceptual economies relative to the more fundamental practice). There is nothing incoherent or unmanageable about the alternative. In fact, it can be shown that a would-be principled account *must* rely on a case-by-case procedure. For, first, although sets and classes are extensionally construed, classification itself – the judgment that what is before us is of this or that kind – is ineliminably intensional. Secondly, any classificatory practice that means to capture or adjust the normative sensibilities of a society must attend to the intensional distinction embedded in that society's linguistic practice. And thirdly the classificatory practices of all natural languages are peculiarly concerned with contextual distinctions – where contexts are themselves intensionally specified, incapable of being identified in any extensionally principled way, and peculiarly centered on paradigms and case-by-case comparison.[8]

The irony is that the usual defense of a principled ethic rests upon its rationality and impartiality. But if language functions as here sketched, then rationality and impartiality must be compatible with a case-by-case procedure – since a principled ethic presupposes it. Now, a principled ethic goes further, in a substantive sense, since it claims that pertinent normative appraisals are arrived at by and only by construing the cases before us as members of pertinent ethical classes. It is just *this* assumption that makes possible the good-conscience sacrifices of applying egalitarian and utilitarian principles; and it is just *that* that the triage conception being developed here contests so forcefully.

What we have shown, therefore, is not merely the coherence of triage (as construed) or the possibility of applying it in a rational way, but also the profound sense in which the phenomena of triage may be taken to challenge the basic orientation of much of Western ethical theory. There can hardly be any doubt that Western ethical theory is strongly committed to the formulation of a principled system producing verdicts. Here, there are two extreme mistakes to be avoided. The first we have already considered: namely, that a seriously proposed theory must be a principled theory. The second is that a seriously proposed theory cannot admit any principled judgments – which, if anything, is even more preposterous. Because of the equivocation remarked, however, it is not entirely clear that anyone has actually maintained the second view, that is, that *any* concession to principled judgments (or the use of well-intentioned procedural regularities even if not meant to be construed in terms of a principled ethic) either constitutes a fatal concession to utilitarianism or confirms the indefensibility of pertinent moral generalizations. But Helmut Thielicke, for one, has certainly been so interpreted:[9]

> Whoever [he says] seeks for such criteria [objective criteria for triage], by that very fact – though perhaps without his even knowing it – has already surrenderd to the utilitarian point of view. He has attempted to apply quantitative standards to that which, because it bears the quality of the unconditional, simply cannot be measured. And in so doing he has missed the very point of the humanum, the alien dignity [which derives from God's having created man] ([28], p. 172).

It is an extravagance, however, to claim that full respect for the human being requires an admission of God's creation; and it is a worse extravagance to suppose that no bona fide moral concerns are best construed in principled terms – for instance, regarding the bare survival of the human race faced with the prospect of nuclear holocaust, or the provision of an ample, ecologically safe world for indefinitely many future generations. It is also a mistake to conflate criteria with principles, and to suppose that principled forms of triage must be utilitarian rather than egalitarian or deontological or of some such alternative sort. But it may well be that Thielicke is actually opposed to principled triage (that he has misrepresented his own view), and that he is not opposed to the moral pertinence of criteria and other general considerations (which he himself occasionally supplies). Joseph Fletcher, by contrast, construes utilitarianism as a way of insuring respect for individuals [9]; and James Childress, who sees utilitarianism as subordinating human dignity to "social role and function," nevertheless holds that egalitarianism does insure "the individual's personal and transcendent dignity" [5]. But Childress cannot then have considered the implications of an egalitarian lottery for triage cases. Sometimes, principled judgments manifest an appropriate measure of respect – and sometimes not.

Now, we are bound to contrast two kinds of principle. Certainly, it would be odd to deny that acting so as to conform with the allegedly unconditional respect that Thielicke claims for man is to act out of principle. But the principle affirmed is, precisely, one that *cannot* (for logical reasons) yield principled judgments; for Thielicke's point is to deny that judgments meant to insure the proper moral respect for persons are addressed merely and only to whoever contingently instantiates some general category of moral concern or appraisal (it being the case that only humans are morally judged). Here, we may borrow – and extend – a particularly useful term coined by Robert Nozick. Nozick calls "a principle of distribution *patterned* if it specifies that a distribution is to vary along with some natural dimension, weighted sum of natural dimensions, or lexicographic ordering of natural dimensions" ([22], p. 156). Let us say more generally that a principle is patterned if it permits only principled judgments. (Nozick's notion then, covers only a certain subset of patterned principles – namely, distributional ones.) In this sense, Thielicke's principle, and of course whatever principle may be said to inform the triage orientation we are developing, are non-patterned principles.

Very likely, the most sustained recent account of the priority and irreducibility of non-patterned principles is that offered by Alan Donagan [7]. Donagan is concerned to isolate a first or fundamental principle of morality that captures the "common morality" of the Judaeo-Christian tradition, including the secular moral philosophy of at least Kant, formulated in terms favorable to natural reason without theological foundations. It is not to be construed merely in terms of contingent *mores;* it is not merely formal; and it preserves a universal, unchanging, substantive law or rule or principle or standard by which, in accord with enlightened reason, we may succeed in fixing the invariant, the "sound and complete," instruction of "common morality" regarding "the duties of human beings to themselves and one another."[10]

Whether Donagan's remarkably sanguine confidence about the determinateness of common morality is justified – say, with respect to suicide, abortion, marriage, divorce, homosexuality and so-called deviant sexuality, war, property, punishment and capital punishment, revolution, future generations, economic scarcity, *and* triage – may well be doubted (particularly given the viability of and evident seriousness with which radically divergent *mores* may actually be supported); but that is not our concern, except perhaps insofar as we thereby acknowledge a certain unavoidable relativism in this most important department of human judgment.[11] To concede relativism of course would be to accommodate rather naturally the very point in dispute between what we have been calling (quite frankly, tendentiously) the triage orientation and its opponent policies within a principled ethic. In any case, in a most subtle and careful way, Donagan offers a variety of formulations of what he takes to be the fundamental first principle of common morality. Two versions, deliberately favoring the Kantian idiom rather than the biblical or theological, are worth mentioning:[12]

> It is impermissible not to respect every human being, oneself or any other, as a rational creature ([7], p. 66). Act so that the fundamental human goods, whether in your own person or in that of another, are promoted as may be possible, and under no circumstances violated ([7], p. 61).

The point to fasten on is Donagan's insistence that, under no circumstances, are the obligations entailed by respect for human beings to be violated. It is essential to Donagan's view that such obligations be both determinate and invariant, though it is important to grasp as well that advocacy of a non-patterned principle of Donagan's sort is compatible with substantive and irresolvable disagreements about such obligations *and* with a relativistic denial that the principle must yield invariant substantive precepts (for instance regarding the ethically sensitive categories tallied earlier in raising the very question of relativism itself). Moreover, it is essential to Donagan's view of the fundamental (non-patterned) principle – as it is, also, for any view of non-patterned principles – that "tragedy is part of human life; and [that] morality goes hand in hand with tragedy" ([7], p. 61). In the context of advancing this thesis, Donagan rehearses a famous Stoic version of the lifeboat dilemma (preserved by Cicero) and its equally famous Victorian counterpart, the case of *Queen v. Dudley and Stephens,* involving cannibalism at sea, which the defendants pleaded was justified on the grounds of necessity.[13] Donagan upholds the court's condemnation of the killing and the eating of an involuntary and helpless victim, invoking a precept falling under his principle: "everybody ought to do what he can to survive; but he ought also to be prepared to die if he must, and he ought not to buy his life at the price of another's" – which was already recognized in the Stoic solution ([7], p. 174), and which of course clarifies the intended severity of the fundamental non-patterned principle Donagan wishes to champion.

But there are complications. Donagan denies that the young boy who was killed was party to an implicit agreement among the crew, to the effect that the apparent survivors of a disaster at sea could act to take the life of some of their number in order to insure the survival of the rest; yet he does concede that a group of spelunkers under not altogether dissimilar circumstances could be said to have entered into such an agreement ([7], p. 177-180).[14] It is pertinent and intriguing to ask whether more general agreements or contracts of this sort could be said to be implicitly in force among human beings – especially in triage cases. Donagan would never admit that they were. John Rawls does not address the question at all. But those who, like Gerald Winslow, would like to apply Rawl's contractarian theory to triage, must somehow suppose (however implausibly) that they are. In any event, triage cases – which Winslow fails to contrast with those of lifeboat dilemmas – *never* involve advantage to the agents who undertake the distribution in question. Such agents might well subscribe to Donagan's linkage of tragedy and morality: they would certainly insist that no solution proffered could fail to involve *the decision not to distribute the resources in question to some human beings who might have been aided by another feasible distribution.* On a patterned principle, no tragedy would be entailed by that distribution; on a non-patterned principle, tragedy could not be denied. Yet both distributions might apply the same criteria of equality.

If, in triage, it is a reasonable requirement that the available resources be distributed to aid some, even though not all could possibly be aided – here, Elizabeth Anscombe convincingly maintains that "[All] can reproach me if I gave [an available, scarce life-saving drug] to none" [1] – then could we ever say with assurance that it was or was not in accord with a non-patterned principle of respect that we deny such drugs *to this one or that?* Donagan does not address the question [7]; and Edmond Cahn's discussion of another well-known lifeboat case, *United States v. Holmes* (which involved a lottery to jettison passengers so as to maximize the safety of those remaining), rejects the lottery, maintaining: "I am driven to conclude that ... they [the survivors] must all wait and die together. For where all have become congeners, pure and simple, no one can save himself by killing another." ([4], p. 71).[15] It may be that on Cahn's view, a contractarian reading of triage *would* entail rejecting Anscombe's judgment. It is certainly clear that being a party directly affected by the life-and-death decision one participates in must color the justice of that decision; but it is not clear that those deciding triage cases are not interested parties (particularly where some criteria of selection rather than others must first be decided[16]); and it is not clear that, even if one is impartial in triage, one has not violated – and cannot fail to violate – some non-patterned principle like Donagan's.

For example, the latest principled account of triage – Winslow's application of Rawls's contractarian theory – cannot, except by fiat, resolve the issue of how triage sacrifices involving life, profound medical disadvantage, suffering and pain, could possibly be vindicated. On Rawls's own ordered principles, "equality in the assignment of basic rights and duties" takes lexical precedence, and "social and economic inequalities ... are [to be allowed as] just only if they result in compensating benefits for everyone and in particular for the least advantaged members of society" ([23], p. 14-15),[17] If the needs that arise in triage contexts are among the most basic, however, then they cannot be compromised (or so it seems); and if we must insure "fair equality of opportunity" with respect to all basic liberties before tradeoffs in inequality can be justified (applying Rawls's so-called "difference principle"), then, once again, they cannot be compromised ([23], pp. 73, 75-8).

Rawls's account, however, does not agree with Donagan's, because the first rests on patterned principles and the second, on non-patterned principles. Rawls's account is simply irrelevant for triage – on its own terms; for triage cannot even be recognized as a genuine problem behind Rawls's famous "veil of ignorance" ([23], p. 12). The only possible application of contractarian theory rests with the rational provision of resources that would bring the condition of those in the "initial position" up from one of dire scarcity (triage) to one of moderate scarcity.[18] But that would simply be to ignore triage dilemmas themselves or to replace them with principled solutions under conditions of scarcity – which, in effect, is what Winslow proposes ([31], Chapter 7). The latter move, however, cannot be a mere application or extension of Rawls's view of justice: it implicitly rejects the independence and lexical ordering of the two principles Rawls himself affirms. Rawls's inability to accommodate the triage problem, then, suggests a fatal weakness in all principled moral systems committed to the adequacy of maximizing utility, rational trade-offs, bargaining with self-interested choices, and the like.

This detour of ours through the moral dilemmas confirms the

distinction with which we first pursued them: the advocate of triage who favors non-patterned principles simply (and of course seriously – perhaps, with a sense of tragedy) supports a distribution that fails to eliminate an evil inhering in the very decision and action supported; whereas the advocate of triage who favors patterned principles simply (and of course seriously – though without a need any longer for a sense of moral tragedy) supports a distribution that, all in all, yields the best or at least an acceptable solution under the circumstances, that is, a verdict, without any taint of evil or a sense of valid responsibility for evil inhering in the very decision and action supported. (Of course, tragedy could still be recovered if restricted to the stinginess of nature and time and cosmic accident; but that would not serve to separate our contending views.) We are at a stalemate, then.

It is a kind of arrogance, no matter how humanely advanced, to pretend that the choice in general between patterned and non-patterned principles – in particular, the preference of one over the other in triage cases – is resolvable by a direct appeal to human nature, human reason, the nature of practical reasoning or of human agency, or the "common morality." There is every reason to believe that both patterned and non-patterned interpretations of human dignity are honorably and impressively convincing to different populations and in different circumstances in spite of the fact that their application yields extensionally distinct solutions to particular problems and intensionally opposed readings of extensionally indistinguishable solutions. If that means that there is no fixed, substantive, determinate moral law, so be it. Or, if the moral law is generous enough to tolerate such divergences, then it cannot be very far from being completely vacuous.

Still, there are stalemates and stalemates. The fact remains that the non-patterned interpretation of triage appears to have been completely or very nearly completely neglected in standard treatments of the subject; and the evidence is that it cannot possibly be a weaker interpretation than the patterned one. That is, very plainly put, there is no known argument that shows that any utilitarian, egalitarian, contractarian, deotological or similar patterned account yields a more plausible or more coherent or more compelling or more correct interpretation of the triage cases than the non-patterned account we have favored – that itself rests on some version of the principle of respect. The upshot is that to honor the argument is to put the dominant moral theories in jeopardy. But it would hardly do, for that reason alone, to dismiss our alternative by a kind of philosophical triage.

Temple University
Philadelphia, Pennsylvania

NOTES

[1] An extremely early American instance is provided in Paul F. Straub, *Medical Service in Campaign: A Handbook for Medical Officers in the Field* [27]. This source is cited in Gerald R. Winslow, *Triage and Justice* ([31], p. 47 ff). I have relied very heavily on Winslow's references, which are very full – even though, as will soon appear, I disagree rather completely with his analysis of triage; also, on Stuart W. Hinds, 'On the Relations of Medical Triage to World Famine: An Historical Survey' [16] – which again is not entirely clear about the range or nature of triage.

[2] This is surely very closely related to the problem once so freshly raised in philosophical circles by J. O. Urmson, 'On Grading' [29].

[3] See Winslow [31], Chapter 2, for a discussion of the application of triage to an anticipated San Francisco earthquake.

[4] This would be denied by Gregory Vlastos, because of his extremely strong interpretation of egalitarian justice [30]; but it would be supported by William Frankena, who contests the need to subscribe to Vlastos's version of such justice [11].

[5] It is in fact quite characteristic of Winslow's account that these and similar sorts of cases are homogenized with those of triage; see Chapter 5. Still his summary of lifeboat ethics is instructive [31].

[6] This may serve to explain the sense in which a well-known exchange between Elisabeth Anscombe and Philippa Foot [1, 10] does not really address the triage problem though it does address other sensitive moral issues.

[7] As, for instance, Winslow seems willing to do; see [31], Chapter 3.

[8] For a more pointed analysis of the logic of classes, see Joseph Margolis, 'The Problem of Similarity: Realism and Nominalism' [19].

[9] See Winslow [31], Chapter 4, Notes 45, 50. The development that follows draws on Winslow's sampled texts but yields rather different conclusions.

[10] See particularly Donagan's discussion in Chapter 1, especially pp. 29-39.

[11] For a discussion of relativism, see Joseph Margolis, 'The Nature and Strategies of Relativism' [20].

[12] This is in the content of Chapter 2. The first version Donagan takes to be "the canonical form" [7].

[13] The details are provided in Chapter 6. Winslow discusses the case briefly, without distinguishing between triage and such dilemmas; cf. [31], Notes to Chapter 5, *passim*.
[14] See Kai Nielsen, 'Against Moral Conservatism' [21].
[15] The case has been discussed by a number of theorists concerned with triage and moral dilemmas.
[16] Winslow is particularly helpful here; see [31], Chapters 4-5.
[17] Cited by Winslow [31].
[18] This is the best construction to put on Winslow's recognition of the limitations of Rawls's account ([31], pp. 120-132). See also Norman Daniels, 'Rights to Health Care and Distributive Justice: Programmatic Worries' [6]. Winslow himself admits that "triage fits uneasily in the framework of contract theory" ([31], p. 166).

BIBLIOGRAPHY

[1] Anscombe, G. E. M.: 1967, 'Who Is Wronged?' *Oxford Review* **5**, 16-17.
[2] *Blakiston's Gould Medical Dictionary:* 1972, Third Edition, McGraw-Hill, New York.
[3] Broad, C. D.: 1930, *Five Types of Ethical Theory*, Routledge and Kegan Paul, London.
[4] Cahn, E.: 1955, *The Moral Decision: Right and Wrong in the Light of American Law*, Indiana University Press, Bloomington.
[5] Childress, J.: 1970, 'Who Shall Live When Not All Can Live?' *Soundings* **43**, 339-362.
[6] Daniels, N.: 1979, 'Rights to Health Care and Distributive Justice: Programmatic Worries', *Journal of Medicine and Philosophy* **4**, 174-191.
[7] Donagan, A.: 1977, *The Theory of Morality,* University of Chicago Press, Chicago.
[8] Engelhardt, H. T., Jr.: 1976, 'Individuals and Communities, Present and Future: Toward a Morality in a Time of Famine', *Soundings* **59**, 70-83.
[9] Fletcher, J.: 1968, 'Donor Nephrectomies and Moral Responsibility', *Journal of the American Medical Women's Association* **23**, 1085-1091.
[10] Foot, Philippa: 1967, 'The Problem of Abortion and the Doctrine of the Double Effect', *Oxford Review* **5**, 5-15.
[11] Frankena, W. K.: 1962, 'The Concepts of Social Justice', in R. B. Brandt (ed.), *Social Justice,* Prentice-Hall, Englewood Cliffs, pp. 1-29.
[12] Hardin, G.: 1972, *Exploring New Ethics for Survival,* Viking Press. New York.
[13] Hardin, G.: 1974, 'Living on a Lifeboat', *Bioscience* **24**, 561-568.
[14] Hardin, G. and Baden, J.: 1977, *Managing the Commons,* W. H. Freeman, San Francisco.
[15] Hardin, G.: 1977, *The Limits of Altruism,* Indiana University Press, Bloomington.
[16] Hinds, S. W.: 1976, 'On the Relations of Medical Triage to World Famine: An Historical Survey', in G. R. Lucas, Jr. and T. W. Ogletree (eds.), *Lifeboat Ethics,* Harper and Row, New York.

[17] Hume, D.: 1751, *An Enquiry Concerning the Principles of Morals,* Section iii.
[18] Margolis, J.: 1971, *Values and Conduct,* Clarendon Press, Oxford, Chapter 8.
[19] Margolis, J.: 1978, 'The Problem of Similarity: Realism and Nominalism', *The Monist* **61**, 384-400.
[20] Margolis, J.: 1983, 'The Nature and Strategies of Relativism', *Mind* **92**, 548-567.
[21] Nielsen, K.: 1972, 'Against Moral Conservatism', *Ethics* **82**, 219-231.
[22] Nozick, R.: 1974, *Anarchy, State and Utopia,* Basic Books, New York.
[23] Rawls, J.: 1971, *A Theory of Justice,* Harvard University Press, Cambridge.
[24] Rescher, N.: 1965, 'The Allocation of Exotic Medical Lifesaving Therapy', *Ethics* **19**, 173-186.
[25] Ross, W. D.: 1939, *Foundations of Ethics,* Clarendon Press, Oxford.
[26] *Stedman's Medical Dictionary*: 1972, Williams and Wilkins, Baltimore.
[27] Straub, P. F.: 1910, *Medical Service in Campaign: A Handbook for Medical Officers in the Field,* P. Blakiston's Son and Co., Philadelphia. Cited in Winslow, G. R.: 1982, *Triage and Justice,* University of California Press, Berkeley.
[28] Thielicke, H.: 1970, 'The Doctor as Judge of Who Shall Live and Who Shall Die', in K. Vaux (ed.), *Who Shall Live?,* Fortress Press, Philadelphia, pp. 146-194.
[29] Urmson, J. O.: 1950, 'On Grading', *Mind* **59**, 145-169.
[30] Vlastos, G.: 1962, 'Justice and Equality', in R. B. Brandt (ed.), *Social Justice,* Prentice-Hall, Englewood Cliffs, pp. 31-72.
[31] Winslow, G. R.: 1982, *Triage and Justice,* University of California Press, Berkeley.

ROBERT M. VEATCH

THE ETHICS OF CRITICAL CARE IN CROSS-CULTURAL PERSPECTIVE

The preliminary documents of the conference for which this paper was prepared outlined two primary clusters of ethical issues in critical medical choices. One cluster centered on problems of "autonomy and paternalism." Insofar as the principle of autonomy is seen as implying liberty rights, that is, the right of self-determination, the right to be let alone, the primary significance for critical care decision-making has been the assertion of the right to refuse such care. The critical care questions in the American debate have been: How can the rational individual express his autonomy by deciding to escape the assault of aggressive but agonizing tinkering by medical professionals desperately trying to overpower nature when the patient would prefer to step back and let nature take its course? How can we decide who really is a rational, autonomous individual? And how can we find autonomous surrogates to make decisions for those who are not competent?

The second cluster raises a different, in some ways opposite, set of questions under the rubric of "justice in the allocation of critical care." How can each member of the community get his or her fair share of the wonders of medical science? How can scarce, lifesaving, critical care be spread equitably among the many who desperately want it?

There can be little doubt that these are the dominant ethical questions of critical care in the United States. There can also be little doubt, however, if we are to believe medical ethics specialists reporting from abroad, that in no other nation have these questions dominated the medical ethics discourse the way they have in this country. Tristram Engelhardt, reporting on a trip to China, for example, found that "the notion that an individual would refuse a needed operation was often dismissed with amazement" [6, 12].

In field studies on the ethics of health care delivery in Cuba my colleagues and I repeatedly asked about decisions whether to provide seriously afflicted infants with intensive medical support. We often found that the question literally could not be understood. At one major obstetrics hospital we asked to see what we called the neonatal intensive care unit and were shown two premature babies in incubators. It soon

J. C. Moskop and L. Kopelman (eds.), Ethics and Critical Care Medicine, 191–206.

became abundantly clear that decisions about high technology intensive care for infants were no problem because that level of exotic care simply was not within reach of the Cuban health care budget.

It is rare, perhaps impossible, to find another country where even one of these two clusters of ethical quandaries has so captured the public and professional imagination. Certainly no other country has managed to elevate these two seemingly conflicting moral dilemmas to the pinnacle of the collective conscience simultaneously the way the United States has. It may be worth, then, attempting to gain some further insight into the moral framework that informs American critical care choices by examining alternative approaches to critical care decision making in cross-cultural perspective.

Background research for this paper confirmed my impression that we have relatively little detailed information on the way critical care choices are made in other countries and virtually no systematic comparative insight. My data are drawn from relatively limited field studies of the health care systems of Sweden, Poland, and Cuba together with two years of experience working within the health care system of Nigeria, supplemented by a sparse published literature. In order to provide a framework for this initial exploration let me propose three theses for debate:

(1) A commitment to an extensive and aggressive program of critical care requires a combination of sub-structural economic conditions and a cultural value framework. In particular, such a program requires substantial wealth combined with an orientation of instrumental activism.
(2) Autonomy will be a significant moral problem in critical care when a society combines substantial resources with an ambivalence toward the aggressive use of technology and an individualism that conflicts with traditional professional paternalism.
(3) Justice will be a significant moral problem in critical care when a society combines a perception of significant resource constraints with an optimism about the potential value of the aggressive use of technology and a communitarianism that acknowledges a realistic social responsibility for the welfare of the less well-off of the community.

These are obviously abstractions which will fit the circumstances of any nation, at best, only as approximations. Nevertheless, I propose that the United States more than any other country possesses the combination of economic and ideological patterns that make autonomy a significant issue in critical care ethics. Not only that, I would like to go on to propose that it is also the country that, more than any other, possesses the combination of these patterns that makes justice a substantial moral controversy. Let me explore the two ethical themes in turn.

I. AUTONOMY AND CRITICAL CARE

The theme of patient autonomy in critical care decisions has surfaced from time to time in many countries even if it has not dominated discussion as it has in the last decade in the United States. It has surfaced most noticeably in the informed consent, right to refuse treatment, and euthanasia debates. In Britain, for example, as early as 1936 a bill was introduced into the House of Lords that would have permitted "voluntary euthanasia" under certain conditions. The Euthanasia Society of England prepared a similar bill which was introduced into Parliament in 1969 ([9], pp. 201-206). In both the emphasis was on the voluntary choice of the patient; he or she would have had to file a declaration requesting that euthanasia be administered "to a mentally responsible patient." The legislation is ambiguous with regard to what kinds of interventions would have been covered. The stopping of heroic resuscitative efforts presumably would have been included, but the language of the draft implies active interventions to hasten death – a procedure more meaningful for the chronically ill than for cases of critical, acute intervention (where simple withholding of treatment would accomplish the desired end). No consideration is given to critical choices for incompetent patients who cannot act autonomously.

More recently England has provided the leadership for the development of hospices, refuges for those wanting to escape critical care. This phenomenon testifies not only to the British tolerance of such decisions, but to its relative suspicion of the wonders to be wrought with modern technological activism [10, 21, 22]. It has also provided the ground for the development of more aggressive movements to permit the choice of more actively terminating a life prolonged in critical illness. The British group EXIT (The Society for the Right to Die with Dignity) proposed the publication of a manual instructing its members on how to commit suicide. The limits on the tolerance of such expressions of autonomy in Britain are seen in the fact the the British Medical Association condemned these plans and, in fact, public pressures were so great that publication plans were suspended [2].[1] By contrast, another group of Britons has moved to the United States, organizing under the name Hemlock, and has published a similar volume with relatively little controversy [14]. Americans, it appears, are not only willing to tolerate a great deal in the name of freedom; they also harbor a certain sympathy for the one who is an activist, who will take matters into his own hands and end it all. It has even been described in one article as the American way of dying [1].

In Switzerland recent debate over critical care decisions has surfaced around the case of Urs Peter Haemmerli, a distinguished Zurich physician charged in 1975 with homicide in conjunction with his decision to withdraw medical support including nourishment from a number of elderly, chronically ill, apparently irreversibly unconscious patients [8]. Though never brought to trial, apparently because of lack of evidence, he was suspended from his position as chief of medicine at his hospital and barred from seeing patients for several months.

Largely as a result of this case, the Swiss Academy of Medical Sciences issued "Guidelines Concerning Assistance to the Dying." These emphasize that for the "discerning" patient, "his will concerning treatment has to be respected, even if it is not in agreement with medical indications" ([24], p. 1169). But in contrast with American court cases (see [15, 16, 17]) and the recent recommendations of the President's Commission for the Study of Ethical Problems in Medicine and Biomedical and Behavioral Research [18], the Swiss document, which is merely the statement of the medical organization rather than an official public position, radically constrains the autonomy of the family in acting as agent for the patient. The family must be listened to, but "from the legal point of view the final decision has to be made by the physician" ([24], p. 1169).

In Sweden a small movement exists to promote support of decisions to withhold or withdraw treatment. In a field interview with Loma Feigenberg of the Karolinska Hospital in Stockholm and one of the leaders of that movement, he stressed the long struggle he has had even to get these issues discussed. His emphasis on rights of patients and his explicit use of the term *contract* between the terminally ill patient and the physician suggests a minority view within Sweden where the dominant concern is still with having the physician as expert do what he thinks is best for the patient [13]. Here, as in Switzerland, to the extent there is any toleration of withdrawal of heroic intervention, the emphasis is often on the unreasonableness of perpetuating treatment in hopeless cases rather than autonomous choice by patients or by parents acting as agents for their wards.

These examples make clear that there are efforts in many countries to support the "right to die" or "death with dignity" movements. There are in fact, over 20 national groups organized into the World Federation of

Right to Die Societies. Still, it is clear that they represent, for the most part, vocal minorities. No other country has anything resembling the American crusade for patient autonomy in critical care decisions where hundreds of cases have been heard in the courts, where every state has considered legislation clarifying the right to make choices regarding treatment refusal (13 states and the District of Columbia having passed such legislation), and where hundreds of thousands of copies of Concern for Dying's "Living Will" have been distributed to assist people in making an autonomous choice regarding terminal critical care refusals. It is worth asking what features of the American culture combine to provide this fascination with the theme of autonomy in critical care decisions.

I am suggesting that a society must exhibit at least three critical features in combination before autonomy becomes a dominant ethical theme in critical care decisions. It must have high enough resource levels that care can be offered that is significantly beyond what many of the community consider reasonable or desirable. There must be significant amounts of what I would call "technological skepticism" – doubt about the benefits of high technological innovations such as hemodialysis machines, cardiopulmonary resuscitation, organ transplants and implants, and radiological and chemical therapies together with enough enthusiasm for these technologies to stimulate a significant number of the community's health professionals to want to use them. And finally there must be an odd combination of autonomy and paternalism. There must be an aggressive liberal individualism emphasizing the central moral importance of autonomy as an ethical imperative together with a tradition of medical paternalism that emphasizes the Hippocratic tradition's principle that the physician should do what he thinks will benefit his patient (even if the patient does not happen to concur). All three of these features – high resources, technological skepticism, and liberal individualism – exist in the United States as nowhere else in the world. At least no other society has all three as dominant characteristics.

(1) *Resources.* Some countries never have the opportunity to have controversy over the violation of patient autonomy in critical care because they simply do not have the resources to permit marginal, expensive, high technology interventions to be considered. Reports about the absence of controversy over the patient's right to refuse treatment have reached the author from not only the poorest of nations but countries like England, China, various African countries, Cuba, and

Poland. In field interviews in Poland, just prior to the current political tensions, for example, one highly placed section head in a major research facility complained of the problems of resource constraints saying "every year it gets worse and worse. We have to close some departments. It is a very difficult problem." In such a setting the problem is normally not one of patients or parents trying to get unwanted treatments stopped; it is one of getting access. At a center doing cardiac surgery for congenital heart disease, one informant said they cannot possibly treat all in need. He said they have to say to some parents of children with complex diseases "We have 200 good children [i.e., good candidates for surgery]. Sorry, your child cannot qualify." When asked about parents who might refuse surgery that the medical staff thought was necessary, he had as much difficulty comprehending the problem as our Cuban informants did.

(2) *Technological skepticism.* In some other countries where resources are not a severe constraint the ethical problems of autonomy in refusing critical care are still not serious because there is such widespread enthusiasm for medical technology that only at the fringes is there any resistance. Sweden is perhaps the clearest example. Dr. Feigenberg's group questioning high technology heroic interventions is a small minority. During field studies in Sweden, I asked whether a hospice-type movement was emerging in Sweden. One administrator's response was that he could think of nothing new. We asked to visit an institution where care for the terminally ill was provided, expressing an interest in seeing facilities that might offer palliative, hospice-type care. We were taken to the oncology department of the hospital and given a thorough tour of a massive collection of radiation equipment including a mid-hallway chest examination of a passing female patient who had recently had what the oncologist thought was a particularly impressive radical mastectomy. When another physician was asked about palliative care, he replied, "This is strictly a treatment place. When we can't do anything more, the patients go home." Although the U.S. is sometimes seen as the epitomy of an activist, interventionist mentality, known, for example, for its high rates of surgery and high numbers of hospital beds per capita, Sweden has more than twice as many hospital beds per capita. Only 18 percent of physicians are in general practice. A British expatriate practicing in a Swedish hospital expressed some contempt for what he took to be the Swedish activist attitudes, saying, "In Sweden all can be cured if only the right doctor is reached."

In some countries autonomy in patient decision making is even less of a problem because lack of adequate resources is combined with an enthusiasm for technology. Cuba is an example. One obstetrician in Cuba told of his efforts to promote more natural childbirth being rejected by his professional colleagues who preferred the more technically controlling interventions. Asked about fetal monitoring, he indicated they limited their intervention to eighteen percent of deliveries, but, he said, "If we had enough monitors we would do more." The right of women to refuse monitoring is a question that simply does not arise.

(3) *Autonomy and Paternalism.* Finally, there are some countries where lack of an aggressive individualism prevents autonomy from becoming a critical problem in high technology medical treatment decisions. A Swedish lay person, but one with doctoral level training in the medical sciences, being interviewed on the question of the extent to which the lay person participates in the medical decisions about cancer therapy said "The whole responsibility is on the doctor in charge. I don't question the decision they make." A nurse, who had previously appeared highly sensitive to debates about patients' rights summarized the situation in Sweden by saying, "In Sweden it is only the physician who has any authority to determine whether the patient is to stay [in the hospital] or not." Then as an afterthought she added, "Yet in one sense the patient has his own right over his life." As a leading Swedish sociologist of medicine put it in an interview, "We emphasize equality and deemphasize freedom." A psychiatrist in a leading Swedish psychiatric institution discussing the absence of consent for electroshock treatments said, "Here in Sweden the doctor takes the decision and can do it against the patient's desire." Later commenting on a patient being forcibly medicated she said, "The patient can refuse to take the drug, but the doctor can overrule. That is a decision you must do within yourself." This stands in contrast to recent legal decisions within the United States giving mental patients circumscribed rights of treatment refusal including in some jurisdictions a specific statutory right to refuse electroshock and prescribing due process requirements in cases where treatment is to be rendered against the wishes of the patient.

Similar reports of acceptance of physician dominance of decision making reached the author as anecdotes from Germany, the Philippines, Nigeria, and Poland as well. In Poland one physician, asked about patient involvement in the decision-making process, said, "Here we have none. I have no time for this." Switzerland is another example. Peter Haemmerli expressed explicitly a contempt for the consent of the patient, which is so closely linked to the principle of autonomy. He said, "Informed consent is something that is really only good for the physician. It is a matter of salesmanship. The physician can persuade the family to consent if he wants to. It is meaningless" (quoted in [8], p. 1275). In the debate over his decision to let the elderly hospital patients die, there has never been any discussion of what the wishes of those patients or their families were or whether they had ever expressed any support or resistance to what he did. The Swiss guidelines are concerned about giving the physician freedom to make critical care choices, but there is no evidence of any concern for familial autonomy. In fact, radically contrasting with the central role given parents and other family members in the U.S.A., the family is relegated to a subsidiary advisory role in decision making for critical treatment choices for patients who are unable to speak for themselves.

As in Switzerland, Swedish physicians are permitted to discontinue treatment if there is no hope of return to consciousness. An initial draft of the National Board's position statement required permission of the family to cease treatment, but that was deleted in the final 1976 version. A distinguished Swedish physician who was familiar with Duff and Campbell's [11] published claim that they bring the parents of infants into any treatment stopping decisions said, "I would not have taken the parents into the situation at all. The decision would have been mine." He concluded by commenting, "In Sweden the doctor is a little pope."

If liberal individualism totally dominated medical decision making, of course, autonomy would cease to be controversial altogether. What is required for controversy is the strange combination of widespread societal commitment to liberal individualism combined with a professional ethic for physicians that has been isolated from that liberal influence. That is precisely what has existed in the United States at least until very recently. American Hippocratic medical ethics has been almost as paternalistic as its continental counterpart floating as an island of benevolent authoritarianism in a sea of autonomy-loving liberalism. That, of course, has changed rapidly in the last decade as American physicians began to absorb the American liberalism of our founding fathers. The most vivid contrast is seen in a country like Cuba, which has not only extremely limited resources, a pervasive technological optimism and a deemphasis on individual autonomy, but one of the most egalitarian, anti-paternalistic physician cadres in the world.

II. JUSTICE AND CRITICAL CARE

To the extent that that Cuban pattern exists, it seems hard to imagine much controversy over autonomous patient participation in critical care decisions, especially as manifested by their refusal of treatment. That same pattern, however, might predictably make our second cluster of ethical controversy much more significant. Problems of justice in the allocation of critical care resources, I have suggested, are likely to emerge in countries whose citizens perceive themselves as having limited resources, harbor great hopes that medical technology offers substantial benefits, and accept a more communitarian view of society's obligation to its least well off citizens.

(1) *Resources.* This means that in some small number of countries blessed with good fortune, justice in resource allocation will not dominate the medical-ethical debate because resources simply are not perceived as being scarce. In an interview with a radiologist at a major teaching hospital in Sweden, in response to a question about whether he ever felt constrained to limit care because of cost considerations, he replied, "I don't think it is so much the cost. We don't have as many cost/benefit [problems] as in the United States." When he was pressed further about whether he feels any constraints on his budget, he indicated that the only example he could think of was that he was asking for another CAT scanner and meeting some resistance from city officials. His local hospital already had six. An official at SPRI, the Swedish planning institute for health care, told me, "In Sweden health care is provided in such abundance that Sweden has not had to develop more sophisticated indicators [for allocating resources], but," he said, "with the economic downturn we will be more interested."

(2) *Technological skepticism.* A second cluster of countries seems to have avoided the controversy over justice in the allocation of scarce medical resources for critical care for a quite different reason. They are not as wealthy as Sweden, but they are not as aggressively committed to instrumental activism either. In England, for example, it is widely believed that chronic hemodialysis cannot be made available to patients over sixty-five. The procedure is expensive; England's resources are severely limited. Even hints of such a policy that have drifted across the Atlantic are enough to have generated consternation, controversy, and condemnation in the United States, yet the British do not seem all that distressed. It is not because Britain is totally lacking in a moral sense of

justice. To the contrary, Britain's commitment to providing health care to all through its National Health Service far exceeds any organized expression of a sense of justice in the United States.

Part of the British restraint may be because there is in reality apparently no formal, explicit policy excluding those above a certain age. Efforts on the part of many people to document such widespread rumors have failed. A more plausible explanation of this lack of British controversy, however, appears to be more from its basic social style that here, as elsewhere, lets the British muddle through. The cornerstone of the British health care system is the kindly, traditional GP, the family oriented, primary care physician who is never overawed by high technology gadgetry. It seems that it is not only that resources are constrained that explains the relative tranquility in Britain in the face of decisions to constrain critical care, but at least as much its skepticism over the benefits such technological innovations will bring [3, 4, 5, 23]. This is the context that has permitted Britain to take the leadership in the growing hospice movement.

China may be another example of a country where an awareness of human finitude has ameliorated potential controversy over critical care resources. Surely China is more desperately constrained in its resources than Britain and at least as committed to a communal sense of responsibility for the welfare of its citizens. Yet the traditional Confucian commitment to balance between man and nature and awareness of the limits of human capacities seems to take the edge off a potentially catastrophic health resource problem [19, 25].

(3) *The Claims of Justice.* There is still a third possible explanation for the relative lack of controversy over justice in critical care resource allocation. Even if a country is constrained in its resources and even if it is committed to be instrumentally active in using human ingenuity to conquer man's medical miseries, it is necessary, or so I argue, for a country to have embedded within its social fabric what I have called a communitarian sense, a commitment that the problems of the needy of the society are not merely unfortunate, but are also basic moral/political problems of unfairness requiring social action.

A country that for various cultural, religious, or historical reasons believes that its obligations to provide health care resources do not extend throughout the society will, to the extent that that attitude pervades the society, not be embroiled in controversy over justice in resource allocation. The problems, to be sure, will be perceived by outsiders – or at least outsiders who have a more richly developed sense of justice. But administrators, health care elites, and government officials in places like Brazil, pre-1959 Cuba, contemporary Chile, South Africa, and historical India could often lie undisturbed by the staggering needs for medical resources chiefly because some quirk of their cultures fails to define these needs as problems of justice. It sees them merely as fate or misfortune.

These countries stand in radical contrast to others with a much more highly developed sense of justice as an end-state, patterned phenomenon, a country such as Sweden, for instance, where even a physician who is among the minority in private practice can say that "Almost everybody believes it is the responsibility of the society to take care of everybody from cradle to tomb" or Poland, where all but a small, relatively well-off minority are covered by national health programs.[2] These countries lacking a full communitarian perspective may be similar to Sweden and England in their lack of controversy, but the grounding for that tranquility is totally different. In Sweden a commitment to a needs-based distributional system combines with a feeling that the resources are adequate to provide for those in need. In England certain critical care interventions deemed life-or-death by Americans are just perceived as not all that critical.

It is the convergence of all three of these features – a feeling of limited resources, a commitment to activism, and a communitarian responsibility for those in need – that seems necessary to produce the controversy.

Before letting the matter rest, this analysis seems to beg for two more questions: first, if the convergence of these three cultural characteristics predicts controversy over the just allocation of critical care resources, how can post-revolutionary Cuba – a country that seems to score high on all three – have avoided exploding and, second, how can it be that the United States has managed simultaneously to erupt with the controversy over justice – with individuals and especially social groups demanding their right of access to life–prolonging critical care – and the controversy over autonomy – with individuals demanding their right of freedom from it.

First, the Cuba problem. It is in many ways the type-case of a country combining limited resources, infatuation with Western technology, and egalitarian communalism. Cuba is a country fascinated with Western technology. Their desires far exceed their capacity to buy, especially with American embargoes making it difficult to get equipment from the U.S. We repeatedly heard of campaigns to assure that procedures such as abortion were done in the hospital. One physician, a head of a local polyclinic, expressed his support of technology by ridiculing the writing of Ivan Illich, the social critic whose writings, such as *Medical Nemesis,* often take an antitechnological turn. After offering elaborate praise for lay participation in the planning of health programs, this physician said with reference to Illich, "Some problems really are technological . . . . You can't vote on an appendicitis diagnosis." Cubans were, in part for these reasons, critical of the Chinese "barefoot" doctor. Cuba's constitution is, to my knowledge, the only one in the world with a formal commitment to a right to health care including free medical and hospital care, free dental care, and public health and vaccination campaigns [7].

Why, then, does debate over the ethics of allocating critical care resources not mushroom beyond the proportions seen in the U.S.? First, the constraints on resources are so severe that certain critical care resources, though attractive trinkets glittering in some vision of a future ideal society, are so far removed from reality that they cannot even get on the agenda for debate. The same problem may account in part for relative lack of controversy in China and Poland and even England. It is comparable to why allocating the artificial heart or long-term psychoanalysis in the United States today is mainly a matter for debate among academics anticipating future ethical controversy over resource allocation.

Second, even in those areas where critical care resources are within the realm of fiscal imagination, controversy is minimized in Cuba by the relative lack of disagreement over the basic implications of the ethics of justice. Approximately 3000 physicians – half of those practicing in Cuba at the time of Castro's revolution – left in an emigration leading to ideological purification. With the training of over 12,000 new physicians since then, all socialized into a needs–based view of just allocation and a remarkably homogeneous lay population sharing in or at least tolerating that ideology, there is not much left to debate. Virtually everyone recognizes the need to reallocate resources to the provinces and away from Havana, to promote massive public health campaigns, and to provide occupational health programs. This combination of homogeneity and resource realism simply neutralizes the cultural factors that would otherwise stimulate controversy over justice in critical care resource allocation.

Finally, the lay population of Cuba, while they have a strongly developed sense of lay participation in socio-economic and political decisions, has not yet developed a very full awareness of the extent to which resource allocation problems are essentially value choices rather than technical decisions. A chief of a department in the Ministry of Public Health had spent some time telling me of the impressive involvement that the lay population has in policy decisions. (The principle of *poder popular* has provided a formal mechanism of community participation in such controversial decisions as the hiring and firing of local polyclinic directors.) Still, when asked about the participation of lay people in resource allocation decisions, he responded, saying, "This is a scientific matter. They [the lay people] are not technicians . . . . Maybe a surgeon will do it; it doesn't matter."

If this explains how Cuba fails to have the predicted controversy, it may also provide a basis for understanding how the United States can be saddled simultaneously with controversy over autonomy and justice. If Cuba and Sweden and, by comparison, even England are small, relatively homogeneous nations, the United States is by contrast incredibly diverse. Of course it is not diversity alone that guarantees controversy. If so, China or the Soviet Union, Belgium or Nigeria, would join the U.S. on the list of high controversy nations. Given the particular kind of pluralism in the U.S., however, there are radical differences in perception regarding each of the variables under consideration. It is not inconsistent to hold that the United States is enormously wealthy (thus permitting extensive, sometimes undesired, high technology critical care) and at the same time acutely aware of the limits on its resources. The poor are keenly aware of economic and social diversity from daily media reports and from the proximity of exotic private hospitals and crude public wards. They are also aware of the attractiveness to them of some care elites struggle to escape.

Not only that, U.S. heterogeneity makes possible the simultaneous presence of an acute technological skepticism that triggers autonomy-based demands for freedom from medical intervention and, in other quarters, an aggressive technological optimism. That optimism leads physicians to insist on fulfilling their self-imposed duty to benefit by inflicting care on unwilling patients. It also leads deprived groups to join with them in demanding the right to critical care resources.

Finally, the U.S. heterogeneity explains the side-by-side survival of such radically different principled ethical commitments. It makes possible the survival of the classical Hippocratic paternalism of professional physician ethics almost completely untouched until the last decade by the dominant secular liberal philosophical commitments to self-determination. It also permits the survival of several radically conflicting theories of justice simultaneously in the same society – the egalitarianism of the Judeo-Christian tradition recently secularized in an attenuated form by Rawlsian theory, the liberal individualism of Adam Smith modernized by Nozickian libertarianism, and the aggregating utilitarianism of the public policy specialists and economists willing to sacrifice the rights of individuals for the welfare of the community.

With that checkered pluralism there is little suprise that a nation whose core ethical rhetoric over and over again plays off liberty and justice for all remains in extensive controversy over the meanings of these commitments for critical care decisions. No other society seems to have put together the fateful combination of characteristics to stimulate all-out debate on either the issues of autonomy or of justice. Other countries have resources perceived either as too limited or too plentiful; they have not managed to mix enough technological skepticism to lead a sizeable group to resist critical care technologies with enough optimism, enough activism, to press providers or deprived groups to value this care; they have not mixed a thorough-going love of self-determination with enough paternalism to encourage Hippocratic physicians to believe they should treat without adequate consent; they have either too minimal a sense of responsibility for the needy or one developed so fully and so homogeneously that its implications are beyond controversy. Looked at cross-culturally, it becomes increasingly clear that the United States has had to go through some remarkable gyrations in order to come up with a combination of moral and socio-economic commitments that make possible controversy simultaneously on the right to have access to critical care and the right to refuse it.

Georgetown University
Washington, D.C.

NOTES

[1] After the English EXIT suspended plans to publish, the Scottish EXIT published its version, distributing it on a members-only basis.

[2] See ([20], p. 4), where it is reported that over 99% of the population is now covered by national insurance. By contrast I was repeatedly told within Poland that from five to fifteen percent were not covered but always assured that this was a well-off group of private entrepreneurs who did not need to be covered.

BIBLIOGRAPHY

[1] 'The American Way of Dying: A Do-It-Yourself Guide, but for Whom?' *Hastings Center Report* **10**, 6 (December 1980), 2.

[2] 'BMA Condemns Suicide Publication', *British Medical Journal* **281**, (September 6, 1980), 691.

[3] Bunker, J. P.: 1970, 'Surgical Manpower: A Comparison of Operations and Surgeons in the United States and in England and Wales', *New England Journal of Medicine* **282**, 135-144.

[4] Bunker, J. P.: 1971, 'Economic Incentives for Social Progress In Medicine', *Pharos* **34**, 20-22.

[5] Bunker, J. P.: 1972, 'Surgical Priorities and the Quality of Medical Care', *Bulletin of the New York Academy of Medicine* **48**, 173-177.

[6] Childress, J.: 1979, 'Reflections on Socialist Ethics', *The Kennedy Institute Quarterly Report* **5**, 13.

[7] *Constitution of the Republic of Cuba,* Article 49.

[8] Culliton, B. J.: 1975, 'The Haemmerli Affair: Is Passive Euthanasia Murder?' *Science* **190**, 1271-1275.

[9] Downing, A. B. (ed.): 1970, *Euthanasia and the Right to Death,* Humanities Press, New York, pp. 201-206.

[10] Doyle, D. (ed.): 1979, *Terminal Care,* Churchill, Livingstone, Edinburgh.

[11] Duff, R. S. and Campbell, A. G. M.: 1973, 'Moral and Ethical Dilemmas in the Special-Care Nursery', *New England Journal of Medicine* **289**, 890-894.

[12] Engelhardt, H. T. Jr.: 1979, 'Bioethical Issues in Contemporary China', *The Kennedy Institute Quarterly Report* **5**, 5.

[13] Feigenberg, Loma: 1980, *Terminal Care: Friendship Contracts with Dying Cancer Patients,* Brunner/Mazel, Inc., New York [Translation of Terminalvard thesis at the Karolinska institutet, Stockholm].

[14] Humphrey, D.: 1982, *Let Me Die Before I Wake: Hemlock's Book of Self-Deliverance for the Dying,* Hemlock, Los Angeles.

[15] *In re Dinnerstein,* 380 N.E. 2d 134 (Mass. App. 1978).

[16] *In re Philip B.,* 156 Cal. Rptr. 48 (1st App. Dist., Division 4, 1979).

[17] *In re Quinlan,* 70 N.J. 10, 355 A. 2d 647 (1976).

[18] President's Commission for the Study of Ethical Problems in Medicine and Biomedical and Behavioral Research: 1983, *Refusal of Life-Sustaining Treatment,* United States Government Printing Office, Washington, D.C..

[19] Reynolds, F. E.: 1979, 'Death: Eastern Thought', in W. T. Reich (ed.), *Encyclopedia of Bioethics,* vol. 1, Free Press, Macmillan Publishing Co., New York, pp. 229-235.

[20] Roemer, M. I., and Roemer, R.: 1978, *Health Manpower in the Socialist System of Poland,* U.S. Department of Health, Education and Welfare, Washington, D.C.

[21] Saunders, C.: 1977, 'Dying They Live: St. Christopher's Hospice', in H. Feifel (ed.), *New Meanings of Death,* McGraw Hill, New York, pp. 153-179.

[22] Saunders, C., Summers, D. H., and Teller, N. (eds.): 1981, *Hospice: The Living Idea,* Edward Arnold, London.

[23] Sedgwick, P.: 1974, 'Medical Individualism', *Hastings Center Studies* **2**, 74-76.

[24] Swiss Academy of Medical Sciences: 1978, 'Guidelines Concerning Assistance to the Dying (Euthanasia)', *Schweiz. med. Wschr.* **108**, 1169-1171. [Official translation,

completed spring 1978, of the 'Richtlinien für die Sterbehilfe,' published in German in *Jahresbericht 1976 der Schweizerischen Akademie der medizinischen Wissenschaften.]*

[25] Unschuld, P. U.: 1979, 'Confucianism', in W. T. Reich (ed.), *Encyclopedia of Bioethics,* vol. 1, Free Press, Macmillan Publ. Co., New York, pp. 200-04.

ROSS KESSEL

TRIAGE: PHILOSOPHICAL AND CROSS-CULTURAL PERSPECTIVES

The essays of Robert Veatch [8] and Joseph Margolis [2] illuminate quite different aspects of what Warren Reich elsewhere in this collection [7] has called the "moral absurdities" of critical medical care. For Veatch, consideration of problems of autonomy and justice in the provision of crisis care provides an opportunity to explore the reasons why these questions have only come to dominate the medical ethics discourse in the United States. For Margolis, the problem of providing crisis care, and particularly the "solution" of doing so by a process of triage, forces him to consider the inevitability of the resultant evil, that some people will be neglected, and to explore ways in which such a result might be justified.

I will begin by considering Margolis's distinction between merely sorting and purposefully neglecting candidates for critical care. I will suggest that Margolis's assertion of the lack of adequate justification for triage decisions in medicine will only be deepened if we extend the discussions to include ordinary as well as extraordinary measures of health care. I will then turn to Robert Veatch's view that the United States is unusual, or even unique, among nations for its overt concern for issues of autonomy and justice. Here I will suggest that the conflicts over critical care observable in the United States are part of the larger debate concerning the delivery of all types of health care, a debate that grows out of the recognition that views with respect to health and health care are diverse and that conflicts between justice as freedom and justice as welfare are inevitable. Finally, I will try briefly to place both Margolis's and Veatch's papers within the framework of the volume as a whole.

I

The power of Margolis's essay arises from its ability to force us to distinguish between merely sorting (Margolis's "market triage") on the one hand and sorting in order to purposefully neglect ("medical triage") on the other. Thus, the sorting of commodities that takes place in market triage is morally neutral (subject only to fairness in the sorting process, in

J. C. Moskop and L. Kopelman (eds.), Ethics and Critical Care Medicine, 207-214.

pricing and in distribution), being designed maximally to reward the seller and, perhaps, to protect the buyer. Medical triage, growing as it does out of a condition of scarcity, has quite another purpose, being designed to sort those needing to be served "with the understanding *that not all those eligible can be served*" ([2], p. 174). It is in this sense that the outcome of conducting a medical triage is, ineluctably, the accompanying evil of purposeful neglect. In stating the issue in these terms, Margolis is dramatically showing us that the use of language that is apparently value-neutral, the language of marketplace sorting, is carried over to the value-laden situation of purposeful neglect. Thus, in military medical triage, we speak of the one who decides who shall receive treatment and who shall not as the "sorting officer," and we call the place in which these decisions are made "the casualty clearing station."

As Margolis points out, the sorting of products or services that we call market triage is on the basis of economic norms. It lacks any necessary element of urgency, or even scarcity, and it is applicable to any and all products or services. Thus sheepskins, coffee beans, sponges or tobacco leaves are "triaged" in the marketplace, and (in the marketplace) our concerns are for fair practice, for due process, and for just distribution. There is no demand that some be neglected, and principled approaches to distributive problems can justifiably be employed to help establish public policy.

In contrast, medical triage is dominated by our sense of urgency, by the disproportionate shortage of a product or service, and by its perceived importance. The situation of medical triage is not merely sorting; rather it is sorting for the purpose of neglecting some of the needy. Medical triage is not merely asking the market question "on what basis ought different qualities of goods be defined, and how ought they be distributed?" Rather it asks "on what basis can we justify giving some of the needy *no* share of a necessary good?" In Margolis's view this evil (that some of the needy are neglected) cannot be eliminated, and a triage policy cannot be constructed that can be justified from universal and substantive moral principles. This is not to suggest that Margolis is advocating neglect, rather he is insisting that we recognize and take responsibility for this inevitable outcome of medical triage. To paraphrase Margolis, medical triage demands the acceptance of tragedy while market triage seeks to deny it.

While medical triage takes origin in military medicine, and is extensionally applied to the metaphorical war on disease, it is worth asking whether or not there are significant differences among varying examples of medical triage. I will cite four: first, medical triage as military triage in the management of battlefield casualties; second, medical triage as the management of medical catastrophes in civilian life; third, triage to assign, and not assign, exotic life-supporting therapies that are scarce and likely to remain so; and fourth, triage to assign ordinary means of health care (such as new antibiotics or vaccines) that while in short supply today are likely to become plentiful. All four of these types of situations meet Margolis's criteria for medical, rather than market, triage. There is an urgent need of an important product or service which is in disproportionately short supply, and for which there is an expectation of access. And yet they differ one from the other in important ways, some of which Margolis makes clear, although without perhaps fully illuminating the slippery slope upon which they lie.

Military triage is carried out by physicians who are members of, and whose services flow to, an institution (the army) with a clearly defined mission to which all parties, presumably, agree. Indeed, the military duty of the physician is only secondarily to the patients arriving at his M.A.S.H. His primary responsibility is to maximize the conduct of war. There might be room here at least to consider whether a Rawlsian might justify military triage by suggesting that the parties could have agreed to put aside even the principle of liberty (let alone the difference principle which Margolis discusses) in the interest of subsequently maximizing liberty.[2]

The situation is significantly different in the case of medical catastrophes in civilian life. Here the physician is not as clearly seen as the servant of the institution and, even if he is so seen, there is likely to be far less agreement as to what the precise institutional mission is or ought to be. Perhaps civilian catastrophic crisis is closest to the life-boat problems discussed by Margolis, although both of his cases involve the taking of life by those wishing to survive.

In the case of the assignment of exotic life-supporting therapies that could rarely or never be made available to all, we seem to be even further from the situation seen in military triage. While the need for access is still, following Margolis, crucial, important and urgent, it lacks the immediacy of either of the two catastrophic situations so far described. It would perhaps, be worthwhile exploring whether or why this difference should matter in a situation that must now, and in the future, neglect substantial numbers of the needy.

Finally, there is the case of assignment to ordinary treatment that is in short supply now but will likely become plentiful in the future. As in the case of extraordinary life-supporting treatments, Margolis's criteria are met, and today's solution demands neglect of the needy. Does the fact that tomorrow's solution does not demand such neglect make this case different?

My purpose in citing these illustrations is to ask whether battlefield triage can be clearly set apart from a wide variety, perhaps most or even all, allocation problems in health care. Will resources for the provision of whatever it is that we choose to identify as health care ever be sufficient? Or, to reverse the question, will our notions of what constitutes health care ever be so modest that conditions of scarcity no longer exist? Put yet another way, would not any and all decisions over the allocation of health care (even including the family physician budgeting that "medically considerable resource," his time, among patients each of whom sees his own case as urgent and expects to have access to that resource) fulfill Margolis's criteria for medical triage?

Margolis's paper, then, makes two major claims applicable to health care in general and crisis care in particular. First, some orderings will exist which necessitate neglect of some of the needy, and these orderings can be distinguished from those that do not so necessitate. Second, the neglect of some of the needy cannot be morally justified, and responsibility for this neglect must be accepted. The danger of Margolis's argument, which is most powerful in the case of battlefield military triage, is that it can be so readily extended to other areas of health care which share many of the same features but where our principal concerns are indeed for fair practice and for justice in pricing and distribution.

II

Although Robert Veatch has long been concerned with distributive issues in health care [9], in the context of crisis care he is asking quite a different sort of question; namely, what is unusual about the intellectual, social or cultural climate of the United States that makes its soil so fruitful for bioethical dispute? Veatch states (correctly I believe) that in the United States debates over the provision and distribution of critical care revolve principally around two questions. First, how can individuals assert control over decisions regarding their own care, and especially how can they escape the assault of unwanted critical care? Second, granted that many individuals want critical care, how can each member of the community be assured of access to a fair share? For Veatch, the former is the issue of autonomy in critical care situations, the latter the issue of justice. Recognizing that critical care programs require substantial wealth and an orientation to technical activism, Veatch proposes that concerns over individual autonomy can be expected to arise when resources for critical care are plentiful (or at least adequate), when there is ambivalence concerning the effectiveness of technical intervention, and when a tradition of individualism exists. In contrast, concerns over justice can be expected to arise when the resources of critical care are perceived as being scarce or inaccessible, where there is optimism about their effectiveness, and where a tradition of social responsibility exists. In this view respect for autonomy counterbalances, or is in tension with, both a duty to beneficence (as paternalism) and a sense of justice, and cultures lacking this strong respect for autonomy are unlikely to debate either the provision, or the acceptance, of extensive personal and public health care. It is Veatch's thesis that outside the United States there is very widespread acceptance of both paternalism and social responsibility, an acceptance unopposed by strong concerns for autonomy. As a consequence, critical care is likely to be provided whenever the necessary economic substructure for its provision exists.

It is difficult to differ with Veatch when he frames the discussion in this manner. Rather than attempt to do so, I will raise (although certainly not answer) two questions. Is Veatch's view amenable to extension beyond the boundaries of critical care? And if it is, cannot the United States' dialectic be viewed as another example of that country's historically unusual acceptance of pluralism?

I would argue that the essential difference between the United States and the remainder of the western world (and its "children") is the acceptance of, indeed the high value placed upon, pluralism. In this light the United States' propensity for bioethical debate stems from its pluralism rather than, or at least as well as, from the nature of the highly technical delivery system of the critical care unit. In the remainder of the western world, society is (at least arguably) less pluralistic, and certainly the older, European societies have a much more clearly developed ideal of national homogeneity. This in turn may have led to a greater acceptance of the idea that the power structure (whether it be government or profession) can appropriately speak on behalf of all.

Within health care this has led to a greater willingness on the part of patients to accept medical paternalism and a lesser necessity to create an ethic of secular pluralism for the medical profession's dealings with either patients or the public.

Examination of some of the examples that Veatch gives reveals that he recognizes that these cross-cultural differences are not limited to highly technical, critical care. As he points out, the development of self-help (or better mutual-support) and euthanasia groups, and of the hospice concept of care for the terminally ill, all of which can be considered low technology enterprises, are issues debated significantly differently in different cultural settings. He might also have pointed to differences in the frequency of patient-initiated suits against doctors, differences in the perceived needs for such instruments as a "Patients' Bill of Rights," or differences in the strength of legislation serving to protect human subjects of research.

Thus rather than being restricted to, or even chiefly concerned with, issues of critical medical care, the debate can be seen within the broader framework of social activism. The United States' embrace of pluralism in medicine has led to the acceptance of individualism as opposed to paternalism in single cases, and as opposed to communitarianism in matters of public policy. If this view is correct it will follow that outside the United States an acceptance of pluralism as an ideal and, consequently, a strong concern for personal autonomy will counterbalance neither the concern to do good nor the concern for justice. Without the counterbalance of a concern for autonomy both will lead to the provision, and acceptance, of a variety of forms of health care.

From this viewpoint the nations that Robert Veatch cites might better be viewed from a cultural-historic rather than a currently technological perspective. Thus, both the United Kingdom and Sweden are hereditary monarchies each with a strong ideal of national identity, including an established national church. Poland, China, and Cuba are, in their different ways, building revolutionary, popular nation-states each with a strong ideal of nationality. It might be instructive in this regard to ask why attitudes in Sweden are so different from those in the United Kingdom, and to look for contrasts between them and the two multilingual and multicultural countries cited by Veatch, Switzerland and India, both of which have strong traditions of pluralism although from very different cultural roots.

Veatch's paper, then, is important in drawing our attention to the cultural framework in which the bioethics debate over critical care occurs in the United States and in pointing out how this framework differs from those found in other countries both developed and developing. While implying that extrapolation both to and from the United States is likely to be difficult, his paper suggests fruitful lines of exploration for the future.

III

In conclusion, it is worth emphasizing that the essays of Veatch and Margolis echo many of the themes presented in other papers in this volume. Thus we have in Margolis's thesis of medical triage as an unjustifiable evil a reprise of the notion of "absurdity as inexplicability" sketched by Warren Reich in his parable of the EmergiMedVan [6]. This approach contrasts with Engelhardt's reliance on "conceptual construals" to provide an ethics of finitude which eliminates the evil of injustice (though not of misfortune) [1]. Veatch notes similarly contrasting views of justice in his transcultural survey. In addition, both Margolis's and Veatch's essays forcefully remind us that not all forms of the medical encounter conform to Edmund Pellegrino's ideal of the doctor as first, last and always the servant of the patient. Many medical decisions are, whether we wish it or not, policy decisions, and Veatch's cultural analysis properly reminds us that, in a pluralistic society, the medical professionals too will likely vary in their views of what constitutes the medical good.

Finally, both Margolis and Veatch, Margolis more starkly perhaps, address the distributional issues raised by the papers of John Moskop [3] and Gregory Pence [4]. How much of our national treasure are we willing to devote to health care, to what kinds of health care should it be devoted, who shall receive care and who shall be neglected, and how shall these decisions be made? In critical care medicine, as in much of modern medicine, these economic questions lie at the heart of the changing nature of medical practice.

The Graduate School
University of Maryland, Baltimore

NOTE

[1] In discussing the justification of conscientious refusal Rawls concludes that military "conscription is permissible if it is demanded for the defense of liberty" and that it is appropriate that "citizens agree to this arrangement as a fair way of sharing the burdens of national defense" ([5], pp. 380-381). Such citizens might also be expected to appreciate that this subjected them to the evils of medical triage.

BIBLIOGRAPHY

[1] Engelhardt, H. T., Jr.: 1985, 'Moral Tensions in Critical Care Medicine: "Absurdities" as Indications of Finitude', in this volume, pp. 23–33.
[2] Margolis, J.: 1985, 'Triage and Critical Care', in this volume, pp. 171–189.
[3] Moskop, J. C.: 1985, 'Allocating Resources Within Health Care: Critical Care vs. Prevention', in this volume, pp. 147–161.
[4] Pence, G.: 1985, 'Report of the President's 2003 Commission on the Fall of Medicine', in this volume, pp. 163–170.
[5] Rawls, J.: 1971, *A Theory of Justice,* Harvard University Press, Cambridge.
[6] Reich, W.: 1985, 'A Movable Medical Crisis', in this volume, pp. 1–10.
[7] Reich, W.: 1985, 'Moral Absurdities in Critical Care Medicine: Commentary on a Parable', in this volume, pp. 23–33.
[8] Veatch, R. M.: 1985, 'The Ethics of Critical Care in Cross-Cultural Perspective', in this volume, pp. 191–205.
[9] Veatch, R. M.: 1976, 'What is a "Just" Health Care Delivery?' in R. M. Veatch and R. Branson (eds.), *Ethics and Health Policy,* Ballinger Publishing Co., Cambridge, pp. 127-153.

STANLEY J. REISER

CRITICAL CARE IN AN HISTORICAL CONTEXT

The struggle to help critically ill patients has challenged and troubled physicians throughout history. This essay examines three periods in which illuminating discussions of this occur – the Hippocratic era of Greek medicine (5th–3rd century B.C.), American medicine of the nineteenth, and the twentieth centuries.

I

Hippocratic physicians were concerned about the question of what constituted the limits of the medical art. They struggled with the issue of instituting appropriate medical actions in the face of serious illness. A treatise in which this matter is faced directly, is called *The Art.* It is written to defend medicine against claims by critics who argued that physicians did little to heal patients many of whom either got better without medical ministrations, or died in spite of them. In response to these critics, the author states clearly the purpose and goals of medicine: "In general terms, it is so to do away with the suffering of the sick, to lessen the violence of their diseases, and to refuse to treat those who are overmastered by their diseases, realizing that, in such cases medicine is powerless" ([10], p. 143).

This striking last phrase justifying refusals to treat is elaborated upon later in the treatise: "For if a man demand from an art a power over what does not belong to the art, or from nature a power over what does not belong to nature, his ignorance is more allied to madness than to lack of knowledge. For in cases where we may have the mastery through the means afforded by a natural constitution or by an art, there we may be craftsmen, but nowhere else. Whenever therefore a man suffers from an ill which is too strong for the means at the disposal of medicine, he surely must not even expect that it can be overcome by medicine" ([10], pp. 203-205).

This perspective on critical illness is formed from several ideas. One is the relation between the power of Nature and the power of the medical

J. C. Moskop and L. Kopelman (eds.), Ethics and Critical Care Medicine, 215-224.

art. It was widely believed then that Nature, a force which encouraged re-establishment of the equilibrium of the biologic forces upset in disease, was more potent than anything man could wield. Hippocratic Greek doctors sought to assist Nature in acting on illness, using moderate therapies as supportive agents. Crucial to successful therapeutics was their timing. A modest assist to Nature, given at a critical period in the course of the illness, could have a great effect. This viewpoint towards the medical art thus encouraged a general moderation in the use of therapy.

A second idea that was significant in this regard is also reflected in the passage quoted from *The Art.* The Greek physician was encouraged in the view that a given therapy had rational limits in relation to some illness. And it followed that those who used therapies in a manner that disregarded their natural scope stained the authority of medicine and committed the sin of hubris. This dual commitment to the power of Nature and to an acknowledging of limits to the power of medical means placed philosophical restraints on excessive intervention in grave disease.

Such ideological motives were reinforced in Greek medicine by two other considerations – one social, the other religious. The life of the physician of this era was not easy. Because it was a manual art, medicine had a relatively low status. The physician was often forced to lead the life of a wanderer to make a living, traveling from place to place to offer service to those who were sick. The need to quickly establish a reputation as a good healer when newly arrived in a town led physicians to seek cases in which the likelihood of recovery was good. Hence taking the case of a critically ill patient posed to such itinerants social risks that some physicians seemed bent on avoiding.

If, for these reasons, patients could not find healers willing to assume the burdens of critical care, both understood that two alternatives remained open – entreating the gods, or suicide. While Hippocratic medicine sought to turn the medical art away from reliance on invocations to deities as an explanation or therapy for illness, they continued to have a place in Greek medicine, as Plutarch notes: "Those who are ill with chronic diseases and do not succeed by the usual remedies and the customary diet turn to purifications and amulets and dreams" ([6], p. 245). Suicide was also an alternative, acceptable in

Greek society for those whose illness was incurable and caused great suffering.

II

A point of view antithetical to that developed about critical care therapy in Greek medicine was propounded by American physicians who practiced the so-called "heroic therapy" in the first half of the nineteenth century.

This therapy was derived from the same humoral view of illness used by the Hippocratic Greeks. It proposed that the body is composed of several basic elements which are in equilibrium during health. Disease altered this balance, and was to be counteracted by therapies aimed at the restoration of an equilibrium or health. The Greeks used modest means – diet, exercise, bleeding in small amounts – to help Nature, the principal healer, do the job. The nineteenth century doctors had a different viewpoint and different remedies.

The scientific revolution of the seventeenth century had created a perspective towards Nature that emphasized domination rather than alliance with it. The belief emerged that science would eventually understand the workings of Nature through an experimental method that analyzed complex structures into their components. From such understanding, the goals of manipulating and dominating Nature were logical outcomes.

Nineteenth century doctors approached illness with this alternative view of Nature, and with a therapeutic armory appropriate to this ethos of domination. They sought to restore balance by vigorous efforts to expel unwanted elements through all possible body outlets – the gastrointestinal tract by emetics like ipecac and cathartics like castor oil; the salivary system through mercurous chloride. People were given agents to produce copious sweating, and drugs to produce pus-generating blisters. But the remedies most widely used, and seemingly most effective, were those that bleed the body – leeches and scarification to produce modest local bleeding, venesection to produce copious bleeding. The tangible evidence of the body ridding itself of fluids and secretions by use of these remedies convinced the doctor the remedies worked, and also the patient, who believed too in the goodness of taking medicine to expel disease, and thereby restore a healthful balance to his body. Two cases illustrate the addiction of patients and doctors to medicines and heroic measures. The patient's viewpoint is found in a story related by Oliver Wendell Holmes, the Boston doctor, in 1860. A colleague was called to treat a patient with a terribly sore mouth: "On inquiry," Holmes writes, "he found that the man had picked up a box of unknown pills, in Howard Street, and proceeded to take them, on general principles, pills being good for people." The pills, it turned out, contained mercury ([9], p. 186; see also [15], p. 9).

The last sickness of George Washington demonstrates the doctor's perpective on "heroic" measures. On December 14, 1799 Washington fell ill with a severe sore throat and difficult breathing. A pint of blood was removed by his overseer, without symptomatic relief. A doctor was summoned who bled another pint of blood and blistered his throat. Consultants soon arrived and removed a quart of Washington's blood. He died that night ([11], p. 7).

III

The approach of medicine in the twentieth century to the critically ill has been influenced fundamentally by the growth of technology and deep concerns about the ethics and costs of its use. The successes and the dilemmas of this development are portrayed particularly well in the use of therapy to forestall respiratory failure.

During the nineteenth century attempts occurred to develop machines that enveloped the body or chest, and produced a pressure within that caused the chest wall to expand and contract. An early ventilator or iron lung of this kind was discussed by a Kentucky doctor, Alfred Jones, in 1864. He claimed it would cure a host of ills, including paralysis, rheumatism, and dyspepsia. A more believable and one of the earliest working iron lungs was introduced in 1876 by Woellez. The operator intermittently decreased pressure about the patient by opening and closing a large bellows attached to the ventilator [7].

By the first quarter of the twentieth century, efforts to create a technological response to respiratory paralysis achieved more success. Philip Drinker and Louis Shaw built and tested in 1927 an "iron lung," which sustained the breathing of a child stricken with poliomyelitis for 122 hours before she succumbed to the illness [5].

As efforts to develop improved mechanical devices to assist ventilation progressed, developments on a different front were organizing medical services to cope with patients experiencing physiological crises as a result of surgical operations, trauma, and war. A three-bed unit to provide care for postoperative neurosurgical patients was put together by W. E. Dandy at Johns Hopkins Hospital in 1923. In 1930 Kirschner, director at the University of Tuebingen's surgical hospital, established a special unit for the care and therapy of patients recovering from surgery, and also the critically ill. It allowed new techniques of care to be applied by the most skillful and experienced doctors and nurses to patients in greatest need of assistance.

World War II promoted the development of units to deal with the severely injured and demonstrated the necessity of whole blood transfusion, early operation, improved postoperative care, and the rapid transport of the injured to medical facilities. The fire in the Cocoanut Grove cocktail lounge in Boston in 1942 also contributed to understanding of critical care. To deal with the large number of seriously burned people, an entire floor of the Massachusetts General Hospital was converted into a trauma unit in which physicians, nurses, and students divided up the tasks of emergency and continuing care.

In 1952, an event occurred which linked the techniques of surgery and post-operative care with the technology of respiratory assistance, and firmly established the place of the intensive care unit in modern medicine. In that year, Denmark bore the brunt of a severe epidemic of poliomyelitis. In the last six months of the year the Hospital for Communicable Diseases in Copenhagen admitted nearly three thousand patients, over 300 of whom had respiratory paralysis requiring ventilatory support. Twenty-seven of the first thirty-one patients with this problem died despite therapy, and all within a period of about three days. As a 12-year old girl was nearing death from the same causes, Dr. Bjorn Ibsen, Senior Anesthetist of the hospital, was called as a consultant to evaluate her and suggest therapy. He immediately performed a tracheotomy, inserted a cuffed endothracheal tube, and manually ventilated the patient in the conventional to-and-fro manner. These interventions forestalled her death and led to her improvement.

The therapeutic ideas enunciated by Ibsen became accepted throughout the region. Patients likely to develop respiratory complications were sent to special wards for observation and recording of vital signs, tracheotomies and cuffed tubes were used to deal with crises, manual ventilation was used in place of or to supplement respirators, and measures to deal with shock were instituted. This led to enhanced cooperation between anesthetists and physicians in treating respiratory failure, with the result that mortality from polio and other respiratory insults began to decline. In 1958 respiratory care units were established in Toronto, and Southampton, England. In the same year multidisciplinary intensive care units opened in Baltimore and Uppsala, demonstrating the spread of the physiologic principles and practices of respiratory care to other areas of medicine [8, 14].

It was an anesthesiologist, Dr. Bruno Haid, of the University of Innsbruck, who raised publicly some of the earliest ethical dilemmas resulting from the success of the new techniques of critical respiratory care. He did so in a letter to Pope Pius XII on the subject of resuscitation. In the practice of anesthetists on patients with head wounds, patients following brain surgery, and patients unconscious because of trauma and anoxia to the brain, moral issues seemed to Dr. Haid more perplexing than those related to physical practices. He was troubled about what to do in cases when the patient's condition became stationary and it was clear that only automatic, artificial respiration was maintaining life. If the family insisted that the doctor remove the apparatus to allow the patient who is "virtually dead" to die in peace, should the doctor comply?

The Pope, in a 1957 paper [13], formulated the moral problems posed by modern techniques of resuscitation in three questions: First, does the physician have a right, or obligation, to use these techniques in all cases, even in those the physician judges to be hopeless? Second, does the doctor have the right, or obligation, to remove the respiratory technology when after several days the deep unconsciousness fails to improve, if death will likely ensue? What role should the family's wishes play in this decision? Third, should a patient like the one above, whose life is sustained only by artificial means, be considered *de facto* or *de jure* dead, or must the circulation have stopped before such a pronouncement can be made?

The Pope's thoughtful response significantly influenced the discourse on appropriate intervention for the critically ill. It still does. He asserted as a basic principle of natural reason and Christian morals, that a sick person, and those charged with the responsibility of overseeing his care, are bound to institute measures to preserve life and health. But he held that this requirement implied only the use of "ordinary means – according to circumstances of persons, places, time and culture – that is to say, means that do not involve any grave burden for oneself or another" ([13], p. 502). A more strict obligation would be too weighty for most to bear, and also divert the concerned parties from the spiritual issues of life-threatening crises. Yet he emphasized one was not "forbidden to take more than the strictly necessary steps to preserve life and health, as long as he does not fail in some more serious duty" ([13], p. 502).

From these ideas, it followed that neither physician nor family were under moral obligation to use modern artificial respiration apparatus in all cases of deep unconsiousness, if their use constituted extraordinary means for either party. Even when its withdrawal produced circulatory arrest, the Pope held that the interruption of resuscitative efforts was only an indirect cause of death.

At the end of his commentary, the Pope grappled with the issue of when is one "dead." Whether death was defined basically by brain or circulatory functions was an important question in dealing with the deeply unconscious person. With reservations about his capability to deal adequately with the matter, the Pope formulated a response: "Human life continues for as long as its vital functions – distinguished from the simple life of organs – manifest themselves spontaneously or even with the help of artificial processes" ([13], p. 504). He thus anticipated the medical efforts to revise the circulation-based view of death, which first occurred in 1966. Then, a Harvard University committee of experts in medicine, law, ethics, and history formulated a new criterion of death around the presence of "irreversible coma" [1]. It was no coincidence that its chairperson, Dr. Henry Beecher, was an anesthesiologist.

The explosion of technologic capability in the 1960's to sustain individuals with severe physiologic deficits helped to generate an intense debate about the ethical problems of treating the critically ill. And much of this debate then, as now, was focused on the matter of how, morally, to limit technologic intervention in life-threatening crises. As this debate continued much emphasis was given to the right of the individuals involved to express a preference about what they would have done to them in this regard. A concern that such preference could not be given when the patient was irreversibly unconscious led in the early 1970's to a movement to establish "living wills" as acceptable forms to state a patient's wishes.

The most well-known legitimization of the living will was the affirmation given in the California Natural Death Act, approved in 1976 [3]. In it, the legislature asserts "that adult persons have the fundamental right to control the decisions relating to the rendering of their own medical care, including the decision to have life-sustaining procedures withheld or withdrawn in instances of a terminal condition." It expresses a concern that the inclination of health professionals to prolong life artificially through the use of modern technology may cause persons with a terminal condition loss of "dignity and unnecessary suffering, while providing nothing medically necessary or beneficial." By recognizing the right of patients to construct a written directive requesting physicians to withhold or withdraw life-giving therapies in the event of a terminal condition, it provided the medical profession with a guide about the lawfulness of honoring the request.

During the 1970's the ethical issues of giving critical care were enlarged by ones bearing on the clinical value of the effort and its economic price. As the pace of technologic innovation in medicine enlarged and its cost rose, a new urgency developed to examine carefully the physiologic benefits and limits of technologic interventions, especially long and expensive procedures such as those associated with critical care. For example, several studies found coronary care units established along the model of respiratory care units in the mid-1960's, did not have a lower mortality rate than home care for heart attack victims in randomly matched patient groups [2, 12]. It has become clear that better prognostic indices are needed, which will allow clinicians to separate accurately those with a critical care problem who will benefit from high technology, high cost interventions, from those who will not. This, in turn, will help meliorate the issue of cost and reduce, but not eliminate, the socially and clinically troubling issue of how to allocate expensive but limited resources among various claimants to them. This matter is not only a problem for the nation at large, but for clinicians who on a daily basis must allocate scarce hospital beds, personnel, and equipment among patients.

What has emerged during the last quarter century in therapy for the critically ill is a view that continuous, comprehensive evaluation of such interventions is required and should involve study of their clinical, ethical, social and economic dimensions. The integration of this constellation of variables has come to be called technology assessment.

It is a testament to their wisdom that, in seeking to resolve the moral and clinical problems of life-sustaining critical care interventions, we can be helped by ideas proposed over two millenia ago by the Hippocratic Greek physicians. They, like us, were concerned with developing a conceptual approach to the question of how to think about the limits of medical intervention. Their concern with the need for rational definitions of what given therapies could accomplish, and their view that overreach in the application of therapy is harmful to the patient, the physician in attendance, and the stature of medicine, reinforce modern efforts to learn more precisely, through technology assessment, where we may technologically intervene with good effect, and where we cannot.

But we too have been innovators. To the credit of contemporary medicine, the view is becoming widespread that when we cannot achieve retrieval of life or functions through technology, the possibility for medical care is not at an end. Rather an alternative plan of precise and directed therapy should be instituted aimed at support rather than cure. Such a view, articulated in the medicine of the past decade (see [4, 16]), represents a significant accomplishment for this generation of practitioners. We now have in place the possibility of balancing the sharp-edged, intensive technologic interventions that save so many from death and disability with those carefully titrated and studied physiologic and psychologic therapies that sustain patients for whom modern medical technology holds no solutions. If we succeed in transmitting this balanced approach to care to medical graduates of the present, ours will be the medical generation remembered for its humanistic and technologic synthesis.

University of Texas Health Science Center
Houston, Texas

BIBLIOGRAPHY

[1] Ad Hoc Committee of the Harvard Medical School to Examine the Definition of Brain Death: 1968, 'A Definition of Irreversible Coma', *Journal of the American Medical Association* **205**, 337-349.

[2] Bloom, B. S. and Peterson, O. L.: 1973, 'End Results, Costs and Productivity of Coronary Care Units', *New England Journal of Medicine* **288**, 72-78.

[3] California Health and Safety Code, Sections 7185 through 7195 (Sept. 30, 1976).

[4] Cassem, N.: 1978, 'Being Honest When Technology Fails', *Harvard Medical Alumni Bulletin* **53**, (October), 32-37.

[5] Drinker, P. and McKhann, C. F.: 1929, 'The Use of a New Apparatus for the Prolonged Administration of Artificial Respiration', *Journal of the American Medical Association* **92**, 1658-60.

[6] Edelstein, L.: 1967, 'Greek Medicine in Its Relation to Religion and Magic', in O. Temkin and C. L. Temkin (eds.), *Ancient Medicine: Selected Papers of Ludwig Edelstein,* The Johns Hopkins Press, Baltimore, pp. 205-246.

[7] Grenvik, A., Eross, B. and Powner, D.: 1980, 'Historical Survey of Mechanical Ventilation', *International Anesthesiology Clinics* **18**: 2, 1-10.

[8] Hilberman, M.: 1975, 'The Evolution of Intensive Care Units', *Critical Care Medicine* **3**, 159-165.

[9] Holmes, O. W.: 1883, 'Currents and Counter-Currents in Medical Science', in O. W. Holmes, *Medical Essays: 1842-1882,* Houghton, Mifflin, Boston, pp. 173-208.

[10] Jones, W. H. S.: 1923, *Hippocrates,* Vol. 2, Harvard University Press, Cambridge, Mass.

[11] Kaufman, M.: 1971, *Homeopathy in America: The Rise and Fall of a Medical Heresy,* The Johns Hopkins Press, Baltimore.

[12] Mather, H. G. *et al.*: 1971, 'Acute Myocardial Infarction: Home and Hospital Treatment', *British Medical Journal* **3**, 334-338.

[13] Pius XII, Pope: 1977, 'The Prolongation of Life', in S. J. Reiser, A. J. Dyck and W. J. Curran (eds.), *Ethics in Medicine: Historical Perspectives and Contemporary Concerns,* MIT Press, Cambridge, pp. 501-504.

[14] Pontoppidian, H. *et al.*: 1977, 'Respiratory Intensive Care', *Anesthesiology* **47**, 96-116.

[15] Rosenberg, C.: 1979, 'The Therapeutic Revolution: Medicine, Meaning and Social Change in Nineteenth-Century America', in M. J. Vogel and C. E. Rosenberg (eds.), *The Therapeutic Revolution: Essays in the Social History of American Medicine,* University of Pennsylvania Press, Philadelphia, pp. 3-25.

[16] Saunders, C. M.: 1972, 'The Care of the Dying Patient and His Family', *Contact,* Summer 1972, pp. 12-18.

PETER C. ENGLISH

COMMENTARY ON STANLEY J. REISER'S 'CRITICAL CARE IN AN HISTORICAL CONTEXT'

The history of critical care is not an overworked topic. As I waited for Dr. Reiser's paper to arrive, I wondered how I would go about writing such a paper, if our roles in this conference had been reversed. This mental exercise created a bit of anxiety largely because so few articles and books on the history of critical care sprang quickly to mind, even fewer that approached the material from an ethical point of view.

Standard indexes in the History of Medicine, the *Wellcome Museum Catalogue* and National Library of Medicine, *Bibliography of the History of Medicine* confirmed that there as yet has been little historical analysis about critical care. 'Critical care' has not achieved the status of a keyword, and I found little by looking under related headings, such as 'emergency', 'triage', 'Red Cross', or 'resuscitation'. Looking for articles on the history of where critical care takes place (ambulance, emergency room, operating room, delivery room, intensive care nursery, intensive care unit, coronary care unit, etc.) also yields surprisingly little. The historical chapter in Paul Durbin's *Guide to the Culture of Science, Technology, and Medicine* [1] contains no reference to critical care, and the index reveals no related citation in other chapters. Ronald Numbers's recent survey of trends in medical history demonstrates that critical care has not been a major area of emphasis in the field [6]. Monographs that might have been expected to comment on critical care issues have largely avoided them. Dr. Reiser's *Medicine and the Reign of Technology* [9] and my *Shock, Physiological Surgery, and George Washington Crile* [4] only touch on these issues, a point to which I will return.

What can a historian contribute to a discussion of the ethics of critical care? If standard histories failed me, the *Oxford English Dictionary* did not, providing one starting point on what is meant by 'critical.' 'Critical' is one of many words whose etymology includes a medical reference. Indeed, several of the non-medical meanings of, 'critical', refer to medical events by analogy. 'Critical' means giving careful judgment based on skill or a body of knowledge. There is a crucial element of timing that in many ways is at the heart of the matter (e.g., a critical time to strike). There is

J. C. Moskop and L. Kopelman (eds.), Ethics and Critical Care Medicine, 225-230.

also uncertainty and risk (a critical time for the economy). In the medical sense, 'critical' specifically means "relating to the crisis or turning point of a disease, determining the issue of a disease." Critical care then is not the same as caring for those who are considered "incurable" or "hopeless." The issue in critical care is yet to be determined; there is risk.

The historian, in my view, might investigate what illnesses or conditions were considered critical at a given time by examining the skills, techniques, and ideas available to the practitioner and by analyzing the physician's notion of timing – when and if to intervene and how a physician decides. How did physicians figure the uncertainty and minimize its risk? How did a physician manage when he had guessed wrong? These elements have changed over time, and perhaps it is here where the future historian will be of help.

The Greek physician is a good topic to begin a discussion of the history of critical care. I am sure that Dr. Reiser would agree with me that the Hippocratic corpus presents no single picture of the Greek physician (nor would we expect a unitary view to emerge from such a collected body of treatises by different authors). The Hippocratic physicians were both willing to tackle critical care and to write about it. Timing, so crucial to the meaning of 'critical', was a major concern for the Greek physician. When to act was determined by knowing the time course of the disease. (Prominent students of ancient medicine credit the Hippocratic writers with the notion that each disease has a specific temporal course.) The time for intervention was in the "critical days" when the turning point would take place.

I disagree with Dr. Reiser, however, that the Greek physician would always keep hands off on critical days; rather, many were decidedly interventionist [8]. Greek surgical texts especially maintain this action-oriented approach. Indeed it is the physician who treated the injured who was most likely to be faced with critical care issues. "On Joints" and "On Fractures" allowed for amputations and relocating joints under certain conditions. "On Wounds of the Head" permitted trephining the skull at once if confronted with a contused wound but never in the face of a depressed fracture [5].

Prominent Hippocratic scholars, such as Owsei Temkin [10] and Ludwig Edelstein [3] have argued that one of the achievements of Greek medicine was knowing when to intervene (the timing element again) and that the Greek physician did not shrink back in the face of critical care decisions. Where I would disagree with Dr. Reiser's analysis and especially with his selection from "The Art,"

... and to refuse to treat those who are overmastered by their disease

is that this passage is really not describing critical illness but terminal or hopeless illness. The issue here is decided; there is no uncertainty; it would be foolish to intervene. I do not wish to enter the debate over which are the "true" Hippocratic texts but rather to suggest that other pictures can be drawn about the relation of the Greek physician and critical care than that of Dr. Reiser.

The Hippocratic texts are just part of a tradition in the ancient world concerned with treating critically ill patients. Homer's *Iliad* contains some 141 wounds and injuries. Two sons of Asclepius, both healers, actively treated the wounds by cleansing, removing foreign bodies, and applying drugs [7].

Celsus's *de Medicina,* Books 7 and 8 (the surgical texts), contains considerable discussion of when and how to intervene (the timing element). The surgeon should treat bowel injuries if the bowel were not ruptured; the surgeon should operate on bladder stone only if the patient were young; the surgeon should amputate a limb in cases of gangrene [2].

I would like to comment briefly on the jump in Dr. Reiser's paper from the Greek period to the 19th century and then to the 20th century. I am sure that Dr. Reiser does not mean to suggest that critical care only occurred in these times, but such an organization reinforces the notion that no really important events occurred in between. I support Dr. Reiser's first sentence that critical care has occurred "throughout history" and therefore regret that he has selected only three examples. Issues in critical care (skill, judgment, timing, uncertainty) clearly are present at all times.

We need careful case studies of critical care, and I would suggest merely two to the future historian in hope of beginning to fill the two millenia between Hippocrates and the 19th century. One might investigate how physicians coped with plague victims from the middle of the 14th century to the early 18th century. What did they learn; how did methods and approaches change? Another potential case study would be wound surgery. I would suggest analyzing the work of four surgeons – Guy de Chauliac (d. 1368), Ambroise Paré (1510-1590), William Cheselden (1688-1750), and Dominique-Jean Larrey (1766-1842). All were generally conservative (indeed this is why surgical historians normally praise them). Yet they did not hold back from the fray. In some instances they were bold operators in the thick of critical care medicine. A potential focus for such a study would be amputation after injury. The issues crucial to the four would include hemorrhage, suppuration, constitutional state of the patient, changing technology of war, what later became known as shock, and where the critical care area should be located (e.g. should a leg be amputated close to the battle lines or further removed). Questions for the historian to grapple with are: the evolution of surgical ideas (which wounds were amenable to a surgical approach), how decisions were made to operate, what prompted or restrained a surgeon, and at what point in the illness did a surgeon operate (timing element). I think such studies would indicate that the concepts regulating notions of critical care were not static.

I now turn briefly to 19th century therapy and critical care. Dr. Reiser has alluded to a body of historical writing that views therapy in the 19th century as progressing from heroic interventionism to hands-off nihilism. I am always surprised in reading textbooks for practitioners how little nihilism there was, [8] even late in the century. I suspect that nihilism was never an attractive alternative for the practitioner when faced with a critically ill patient and his family (as contrasted with the terminally or hopelessly ill patient).

Nineteenth century therapy was certainly not monolithic. In certain areas physicians became decidedly more interventionistic. In diphtheria, physicians introduced tracheostomy in mid-century, intubation in 1885, and later anti-toxin. Issues of timing were always discussed (how sick before operating, when to give anti-toxin, etc.). In surgery the trend in the late 19th century was from conservative to radical in the total number of operations performed, the number of different procedures, the number of diseases considered surgical, and the number of bodily cavities entered.

Dr. Reiser ties the exploding implementation of technology in the 20th century with the growing concern for ethical issues (limits of therapy, "overreach," responsibility, viability, who decides, etc.). A historian might also ask how such decisions were made in the past. My contention is that these ethical issues have always been entwined with critical care medicine but that their resolution may have taken place in a different context than represented by today's conference. To outline this point, I will briefly examine the shift from conservative to radical surgery that followed the introduction of anesthesia and Listerism, roughly after 1880 [4].

Surgical radicalism ushered in an exuberance for operating that led initially to increased morbidity and mortality. In surgery of the thyroid gland, for example, too much cutting could lead to injury of the parathyroids or might not leave sufficient functioning thyroid gland. Operating on the chest led to problems because of the negative intrathoracic pressure. Shock, an unfortunate side effect of radical surgery, occurred in most if not all major procedures. The problem was that surgeons were intervening in new areas of critical care medicine and making things worse.

It is possible to detail a response by surgeons to radical excess, a response that was motivated by moral concerns, but in which the words 'moral' and 'ethical' were never used. Beginning in the last years of the 19th century, a movement sprang up within surgical circles, called "physiological surgery," that sought to cleanse surgery of radicalism. Surgeons such as Harvey Cushing and George Crile took responsibility for surgery's shortcomings, clearly stating the problem, criticizing the zealots, and facing the consequences. Through large hospital case series and laboratory experimentation, this new breed of surgeon altered practice in a myriad of ways to make operations safer for the patient. They experimented with timing of the procedure (always a concern for critical care), technique, anesthesia, length of operating time, and types of anastomoses. What emerged was a consensus "operation of choice" for a given surgical condition. In addition, physiological surgeons improved means of diagnosis, prepared patients better pre-operatively (both physiologically and psychologically), monitored patients carefully during the operation, and developed elaborate post-operative care. Surgical residencies changed how a physician became qualified to practice, and the American College of Surgeons set new standards for hospital accreditation. Surgery and critical care were profoundly different in 1920 than a generation earlier. The physiological surgeons had moderated the excesses in the short run while paving the way for further advances in critical care. Yet each step in the process still awaits careful historical treatment.

In sum, I think we in history need to get to work and to enter and explore the exciting field of critical care.

Duke University
Durham, North Carolina

BIBLIOGRAPHY

[1] Brieger, G. H.: 1979, 'History of Medicine', in Paul T. Durbin (ed.), *A Guide to the Culture of Science, Technology, and Medicine,* Free Press, Macmillan Publishing Co., New York, pp. 121-194.

[2] Celsus, A. C.: 1938, *De Medicina,* Vol. 3, W. G. Spencer (trans.), Harvard University Press, Cambridge (Originally written in first century, first Spencer edition, 1935).

[3] Edelstein, L.: 1967, *Ancient Medicine,* O. Temkin and C. L. Temkin (eds.), Johns Hopkins Press, Baltimore.

[4] English, P. C.: 1980, *Shock, Physiological Surgery, and George Washington Crile: Medical Innovation in the Progressive Era,* Greenwood Press, Westport, Connecticut.

[5] *Hippocrates,* Vol. 3, E. T. Withington (ed.), Loeb Series, Harvard University Press, Cambridge, 1928.

[6] Numbers, R. L.: 1982, 'The History of American Medicine: A Field in Ferment', *Reviews in American History* **10**, 245-263.

[7] Sigerist, H.: 1961, *History of Medicine,* Vol. 2, *Early Greek, Hindu and Persian,* Cambridge University Press, New York.

[8] Reiser, S. J.: 1985, 'Critical Care in an Historical Context', in this volume, pp. 215–224.

[9] Reiser, S. J.: 1978, *Medicine and the Reign of Technology,* Cambridge University Press, Cambridge.

[10] Temkin, O.: 1953, 'Greek Medicine as Science and Craft', *Isis* **44**, 213-225.

NOTES ON CONTRIBUTORS

H. Tristram Engelhardt, Jr., Ph.D., M.D., is Professor, Center for Ethics, Medicine and Public Issues, Baylor College of Medicine, Houston, Texas.

Peter C. English, M.D., Ph.D., is Assistant Professor of History, Assistant Professor of Pediatrics, Duke University, Durham, North Carolina.

Sally A. Gadow, Ph.D., is Associate Professor, Institute for the Medical Humanities, The University of Texas Medical Branch, Galveston, Texas.

Jay Katz, M.D., is John A. Garver Professor of Law and Psychoanalysis, Yale Law School, New Haven, Connecticut.

Ross Kessel, Ph.D., is Professor of Microbiology and Dean of the Graduate School, University of Maryland School of Medicine, Baltimore, Maryland.

Loretta Kopelman, Ph.D., is Professor of Humanities, Chair, Department of Medical Humanities, School of Medicine, East Carolina University, Greenville, North Carolina.

Joseph Margolis, Ph.D., is Professor of Philosophy, Temple University, Philadelphia, Pennsylvania.

John C. Moskop, Ph.D., is Associate Professor of Humanities, School of Medicine, East Carolina University, Greenville, North Carolina.

Edmund D. Pellegrino, M.D., is Director, Kennedy Institute of Ethics, and John Carroll, Professor of Medicine and Medical Humanities, Georgetown University, Washington, D.C.

Gregory E. Pence, Ph.D., is Associate Professor, Department of Philosophy and School of Medicine, University of Alabama in Birmingham, Birmingham, Alabama.

James M. Perrin, M.D., is Assistant Professor of Pediatrics and Director, Division of General Pediatrics, Vanderbilt University School of Medicine, and Senior Research Associate, Vanderbilt Institute for Public Policy Studies, Nashville, Tennessee.

Warren Thomas Reich, S.T.D., is Professor of Bioethics, Department of Community and Family Medicine and Kennedy Institute of Ethics, and Director, Division of Health and Humanities, Georgetown University School of Medicine, Washington, D.C.

Stanley J. Reiser, M.D., Ph.D., is Professor of Humanities and Technology in Health Care, The University of Texas Health Science Center at Houston, Houston, Texas.

Stuart F. Spicker, Ph.D., is Professor of Community Medicine and Health Care (Philosophy), Division of Humanistic Studies in Medicine, School of Medicine, University of Connecticut Health Center, Farmington, Connecticut.

Robert M. Veatch, Ph.D., is Professor of Medical Ethics, Kennedy Institute of Ethics, Georgetown University, Washington, D.C.

INDEX

absurdity (*see* moral absurdity)
Adam and Eve 66
After Virtue (MacIntyre, A.) 168
Aid to Families with Dependent Children (AFDC) xvii, 164
altruism (*see* compassion)
American College of Surgeons 229
American Medical Association–1980 Code 84
Anscombe, Elizabeth 184
The Art (Hippocrates) 215, 216, 226–227
Aristotle 90, 91, 92, 133, 136
  doctrine of the good 133–136
  Ethics 118, 134
  teleology of 133–134
Asclepius, sons of 227
authority (*see* moral authority)
autonomy xiii, xvi, xviii–xix, 8, 13, 16–17 36–37, 44, 49, 54–64, 70–73, 125–126, 130, 141, 143–144, 191–198, 207, 211–212
  in Britain 193
  family 194,198
  hospital 9, 12
  Kant's conception of 45
  national 14, 18
  personal 16, 17, 133
  physician 11–14, 16
  principle of 73, 75, 198
  psychological 54–61, 73
  of regulatory agents 13
  rights 14, 18
  traditional model of 17

Beacher, Henry 221
Belgium 203
Bendixen, H. H. 156
beneficence 79–81, 87, 90, 91, 92, 101, 133, 211
Big Brother 18
biomedical good (*see* good, biomedical)
British Medical Association 193
British Ministry of Health 172
Bobbitt, P. 149
Bondy, P. K. 96, 97
Brazil 201
Brody, Baruch A. 28, 29
Burt, Robert 51, 63

Cahn, Edmond 184
Calabresi, Guido 149
California Natural Death Act 222
cardiopulmonary arrest 122–123
cardiopulmonary resuscitation (CPR) xvi, 122–123, 127–129, 139–144
Carlson, R. J. 156
Carter, Jimmy 164
charity 91 (*see also* compassion)
Cheselden, William 227
Children
  Aid to Families with Dependent Children (AFDC) xvii, 164–166
  Bruton's type agammaglobulinemia 107–108
  Crippled Children's Services (CCS) xv, 108–111, 113, 114
    models of 111–115
  chronic illnesses of 105–116
  intensive care 156
  infants, handicapped 95–96
  respirator dependent infants 109–110
Chauliac, Guy de 227
Childress, James 181
Chile 201
China 191,192, 195, 197, 200, 202, 212
Cicero 183
Cocoanut Grove 219
communitarian sense 200–203
compassion 90, 91–94, 96–97
consent
  informed 8, 31, 41–43, 46–66, 70–76, 86, 87, 119, 121, 123–129, 135, 191–196, 197–198, 204
  refusal of 48–66, 71, 74, 118, 119, 131,

191, 196, 198, 199
third party 84, 132
CPR (*see* cardiopulmonary resuscitation)
Crile, George 228
Crippled Children's Service xv, 108–111, 113, 114
critical care medicine *passim*
absurdities in xii, 11–39
agents of 17
challenges to 154–155
decision making 191, 192–198
distribution of 17–19, 23–24, 105–116, 147–162, 191–206, 207–214
effectiveness of 154–156
symbolic value of 149–150
urgency of 152–154
Cuba 191, 192, 195, 197, 198, 199, 201–203, 212, 213
Cushing, Harvey 228

Dandy, W. E. 219
Decision making, medical 171–178, 191–198
for critical care 191, 192–195
Del Guercio, Louis 147, 148, 150, 151, 167
Denmark 219
Department of Health and Human Services 95–96, 99
Diagnosis-Related Groups (DRGs) 147
diphtheria 228
District of Columbia 195
Donagan, Alan 182–184
Drinker, Philip 218
Dubos, Rene 152
Durbin, Paul 225

egalitarianism xvii, 174, 175, 178, 179, 200–203
in emergency medicine 18, 19
empathy 90, 91
Engelhardt, H. Tristram, Jr. xii, 23–34, 35, 36, 37, 39, 191
England (Britain) 24, 29, 32, 193, 195, 199, 200
English, Peter xix, xx, 225–230
entitlement myth 97–99
Ethics (Aristotle) 118, 134
ethics *passim*
clinical 48
corporate 12–16
of compassiion 91–94, 96–97
lifeboat 175, 184–185
Euthanasia Society of England 193
Evtushenko, Evgeny 91
EXIT (The Society for the Right to Die with Dignity) 193

fiduciary relationship 150–151
Fletcher, Joseph 181
Freedman, Benjamin 152–154
freedom 31, 125–126, 157–158 (*see also* autonomy)
Freud, Sigmund 93
Freudenberg, N. 157
friendship 90, 91, 92
Feigenberg, Loma 194, 196

Gadow, Sally xvi, 139–145
Genesis 66
Germany 197
God 127
acts of 25, 149
grace of 26,27
law of 120
Goldblatt, Ann Dudley 49
good
Aristotle's doctrine of 133–136
biomedical 119, 121–123, 132, 135–136, 140, 141–143
of the patient 117–144
patient's perception of 119, 123–125
social 128–129
techno-medical 119, 121–123, 126, 127, 132, 133
ultimate 119–120, 130–131, 135
Good Samaritan 92
policy 31
program xi, 6
Good Samaritanism 4, 17
gratitude 91
Greek medicine (see Hippocratic)
Griffin, Andrew 156
Guerrero v. Copper Queen Hospital 18, 29
"Guidelines Concerning Assistance to the Dying" 194

Haemmerli, Urs Peter 194, 197
Haid, Bruno 220
Hardie, W.F.R. 134
Hardin, Garrett 175
health care systems
  China 191, 192, 195, 197, 200, 202, 212
  Cuba 191, 192, 195, 197, 198, 199, 201–203, 212, 213
  England, Britain, 193, 195, 199, 200
  Poland 192, 195–197, 200, 202, 212
  Sweden 192, 194, 196, 197, 198, 199, 200, 203, 212
  Switzerland 194, 197, 198, 200
  United States 71, 147, 191, 192, 193, 195, 197, 198, 199, 201–203, 210–212
Hemlock 193
Herodotus 136
Hinds, Stuart 172
Hippocratic
  corpus 41, 84, 85, 226, 227
  era xix, 215–217
  medicine 215–217, 223
  norm or moral rule xiv, 79–101, 198, 204
    collective 80, 100
    individual 80, 100
  oath 80, 85, 105, 109
  physicians 204, 215, 223, 226
  tradition 80–101, 195
history of medicine 215–229
Holmes, Oliver Wendell 218
Hospital for Communicable Diseases (Copenhagen) 219
hospice 193
House of Lords 193
Hume, David 79, 90, 91, 92, 93, 94

Ibsen, Bjorn 219
Iliad (Homer) 227
Illich, Ivan 17, 154–155, 202
impartiality 82, 91, 93–97, 100, 180
India 201, 212
individualism 198–203
inequalities
  in entitlements 25, 28
  natural lottery 24, 28, 29
  social lottery 25, 28, 29
informed consent (*see* consent, informed)
insurers 12
In re Dinnerstein 124
integrity, professional 16

Jackson, D. L. 141
Jehovah's Witness 86
Johns Hopkins Hospital 219
Judeo–Christian tradition 182, 204
justice 23–32, 79–101, 112, *passim*
  and health care xiv–xv, xviii, 79–82, 93–96, 99–101, 128–129, 167–169
  and childhood diseases 110–115
  and critical care 17–19, 30–31, 36–37, 79, 81–83, 99–100, 147–158, 163–169, 191–204
  resources and 80, 95–96, 196, 199, 201
  technology and 196, 199, 201–203
  triage and 171–187, 207–210

Katz, Jay xii, xiii, xiv, 41–66, 69–76
Kant, Immanuel 43, 45, 182
Karolinska Hospital 194
Kessel, Ross xviii, xix, 207–214
kindness 91
Kirschner 219
Kopelman, Loretta xiv, 79–104, 105

Ladd, John 43
Larrey, Dominque–Jean 227
Lavelle, Louis 125
liberalism 168, 198
libertarianism 26–27, 36, 133, 134, 203
lifeboat dilemmas 175, 184–185
Listerism 228
"Living Will" 195
Locke, John 27, 98
lottery 172–173, 175, 176, 182–184
  natural 24–25
  social 25

MacIntyre, Alasdair 168
Malthus, Thomas 175
McKeown, Thomas 154–157
Margolis, Joseph xvii–xix, 171–190, 207–210, 213
Marx, Karl 61
Massachusetts General Hospital 219
Medicaid 13, 29, 108, 110, 111, 147
Medical Nemesis 155, 202
medical–industrial complex 12
Medicare 13, 29, 147

Medicina, de (Celsus) 227
Medicine and the Reign of Technology 225
medicine
  Hippocratic 215–217, 223, 226–227
  nineteenth century 217–218, 227, 228
  twentieth century 218–223, *passim*
Mill, John Stuart 43
Montaigne, Michel de 14
moral absurdity 11–19, 23–24, 32, 36–37, 39, 207, 213
moral authority 25, 27
moral frameworks 90–95
moral principles 44, 112
moral skepticism 15
moral vision, loss of 14–15
Moskop, John xvi, 147–162, 163, 167, 213
Mothers Against Drunken Drivers (MADD) 150
MRFIT study 157, 158
muscular dystrophy 109

Natanson v. Kline 41
National Health Service 200
National Library of Medicine 225
Nigeria 192, 197, 203
1984 (see Orwell, George)
Noble savage mythology 98
"Non–Discrimination on the Basis of Handicap..." 95–96
Nozick, Robert 168, 182, 204
Numbers, Ronald 225

Ogilvie, W. 28
Orwell, George xii, 11, 14–15, 32, 39

Paine, Thomas 28
Pare, Ambroise 227
partiality 80, 82
paternalism xiv, 70,72, 74, 79–87, 100, 101 105, 130 133, 139, 140, 144, 191, 211 212
  Hippocratic 203, 204
  in medicine 195, 197
  social 129
  strong and weak 86
patients' rights (*see* autonomy, consent, freedom, justice, self–determination)

Pellegrino, Edmund xv–xvi, 117–138, 139–144, 150, 213
Pence, Gregory xvii, 158, 163–170, 213
Perl, Mark 74–75
Perrin, James xiv–xv, 105–116
Perry, Ralph Barton 135
Philippines 197
Plutarch 216
Poland 192, 195, 197, 198, 200, 202, 212
poliomyelitis 218–220
power 11–14
President's Commission for the Study of Ethical Problems in Medicine and Bromedical and Behavioral Research 96, 118, 194
preventive medicine 157–158
Professional Standard Review Organi–zations (PSRO) 13

Queen v. Dudley and Stephens 183

Ramsey, Paul 132
Rawls, John 86, 116, 167–168, 184, 185, 204
Reagan, Ronald 164
Reich, Warren xii, 1–22, 23, 24, 30, 31, 32, 35–40, 207, 213
Reiser, Stanley J. xi, xix, xx, 215–224, 225–228
Relman, Arnold 12
Rescher, Nicholas 178
rights (*see* autonomy, consent, freedom, justice, self–determination)
Rockefeller Foundation 13
Ross, W.D. 178
Rousseau, J.J. 98

Schroeder, Justice 41
self–determination xiii, 41, 70–71, 120, 191, 203, 204 (*see also* autonomy, freedom)
Shaw, Louis 218
Shelp, Earl E. 74–75
Shock, Physiological Surgery and George Washington Crile 225
Siegler, Mark 47–53, 64, 139–140, 142
Smith, Adam 204

social security xvii, 164
Society of Friends 92
South Africa 201
Spicker, Stuart xiii–xiv, 69–78
spina bifida 113
Starr, Paul 11
Stockholm 194
Strauss, Leo 133
suicide 216
Sweden 192, 194, 196, 197, 198, 199, 200, 203, 212
Swiss Academy of Medical Science 194
Switzerland 194, 197, 198, 212
sympathy 91

technological scepticism 196–197, 199–200
techno–medical good (*see* good, techno–medical)
Thielicke, Helmut 181, 182
Thomasma, David 156
Temkin, Owsei 226
Toulmin, Stephen 43–46, 72
triage 151, 171–187, 207–210
  coffee 171–172, 174
  justification of 171–187, 207–210
  market xvii, 172–173, 207
  medical xvii, 172–176, 178–179, 184, 185, 207–210, 213
  military 171–174, 179, 209–210
trust 151–152

Ulrich, L. 74
United Kingdom (*see* England)
United States 71, 147, 191, 192, 193, 195, 197, 198, 199, 201–203, 210, 211, 212
United States v. Holmes 184
utilitarianism xvii, 177, 179, 181

Van den Berg, J.H. 70
Veatch, Henry 133
Veatch, Robert xviii, 84, 191–206, 207, 208, 210–213
Veterans Administration (VA) xvii, 164

Washington, George 218
Weil, Simon 90, 93
White, Nicholas 134, 135
Wilmington General Hospital v. Manlove 18, 19, 29
Woellez 218
Winslow, Gerald 184, 185
World Federation of Right to Die Societies 194, 195

Youngner, Stuart 141